Ore Geology

Edited by Miloš René

Published in London, United Kingdom

Ore Geology
http://dx.doi.org/10.5772/intechopen.1005551
Edited by Miloš René

Contributors
Christoph Gauert, Damien Gaboury, Jean-Claude Ngaruye, Miloš René, Richard Sack

First published in London, United Kingdom, 2025 by IntechOpen
IntechOpen is the global imprint of INTECHOPEN LIMITED, registered in England and Wales, registration number: 11086078, 167-169 Great Portland Street, London, W1W 5PF, United Kingdom

For EU product safety concerns: IN TECH d.o.o., Prolaz Marije Krucifikse Kozulić 3, 51000 Rijeka, Croatia, info@intechopen.com or visit our website at intechopen.com.

British Library Cataloguing-in-Publication Data
A catalogue record for this book is available from the British Library

Ore Geology
Edited by Miloš René
p. cm.

This title is part of the Earth Sciences Book Series, Volume 6
Series Editor: Maurizio Lazzari

Print ISBN 978-1-83634-263-2
Online ISBN 978-1-83634-262-5
eBook (PDF) ISBN 978-1-83634-264-9
ISSN 3049-8848

IntechOpen Book Series

Earth Sciences

Volume 6

Aims and Scope of the Series

The world of Earth Sciences, considering the interactions within the geosphere and between the geosphere–biosphere, is a place in which a large number of scientists find and have found over time their own relationship or sector of application precisely because it is absolutely transversal to many disciplines and subdisciplines, which do not necessarily fall within the Geosciences. The objective of this book series is to welcome original scientific contributions both in consolidated contexts and in new frontiers of research, as well as review papers included in the various disciplines of Earth Sciences, but above all, those that show a modern and transversal vision of applications and impacts on the community in a particular historical context, which, following the COVID-19 pandemic, has shifted global attention to sectors that were previously more neglected. In particular, those of mining research and fossil and renewable energy sources, environmental geology and the sustainable use of natural resources and impacts on the built environment, land consumption, geoarchaeology, forensic geology, geotourism/geoheritage, georisks and climate and environmental changes, considered at different scales, up to new applications of geostatistical and geospatial analysis, GIS and artificial intelligence for the definition of forecasting models and scenarios in various sectors of basic and applied research.

Meet the Series Editor

Dr. Maurizio Lazzari has a Ph.D. in Earth Science and is a researcher at the Italian National Research Council, Institute of Cultural Heritage Sciences. Since 2001, Dr. Lazzari has been a Professor of Pedology at the University of Basilicata (Italy) and a geoarchaeologist at the University of Salento (Italy). His research activities are focused on natural and anthropic hazards and risk factors, aimed at safeguarding and conserving settlements and the historical–monumental heritage of the Mediterranean, with particular attention to landslide processes, susceptibility maps, monitoring, and modelling. Since 2004, he has been working as a scientific coordinator for several national research projects studying landslides and triggering factors, natural and anthropogenic risks, geological and geomorphological mapping, soil erosion, preservation of historical and archaeological sites, enhancement of degraded areas, geo-touristic use, and the protection of the landscapes. He is the author of about 150 scientific publications in national and international journals, monographs, book chapters, and conference proceedings concerning applied geology, geomorphology, dynamics of artificial reservoirs, soil erosion, landslides, geoarchaeology, hydrogeological instability, natural hazards, monitoring, cultural landscape, UNESCO Heritage, geoarchaeology, and geo-tourism.

Meet the Volume Editor

Dr. Miloš René, Ph.D., is a scientist at the Institute of Rock Structure and Mechanics at the Academy of Sciences of the Czech Republic. He obtained his MSc in Mineralogy and his Ph.D. in Economic Geology from Charles University in Prague, Czech Republic, in 1968 and 1981, respectively. He has published about eighty-five scientific papers in peer-reviewed journals in the fields of mineralogy, petrology, geochemistry, and economic geology. His recent scientific activity is concentrated on petrology and geochemistry of granitic rocks, mineralogy of accessory minerals, and mineralogy of uranium deposits, especially from the Central European Variscan belt.

Contents

Preface

Ore geology is a significant branch of geoscience that deals with geology, mineral compositions, and geochemical evolution of different ore deposits in the earth's crust. The ore deposits occur in highly variable rock series. In ores, different valuable minerals, which contain metals, are concentrated. *Ore Geology* includes five chapters dealing with the ore geology of tantalum-niobium, gold, polymetallic (Cu-Ag-Zn-Fe-Sb-As-S) and uranium ore deposits.

Chapter 1, "Introductory Chapter: Ore Geology – a Significant Part of Geosciences" by Miloš René, explains ore geology's position in geosciences.

Chapter 2, "Geochemical Characterization of Cassiterite, Ferberite and Columbite-Tantalite from Rwinkwavu, Bugarura-Kuluti and Musha-Ntunga Mineral Districts of the South-Eastern Rwanda" by Jean-Claude Ngaruye and Christoph Gauert, focuses on detailed mineralogy of columbite-tantalite, cassiterite and wolframite from some ore deposits in the South-Eastern Rwanda.

Chapter 3, "Volatile Composition of Fluid Inclusions in Gold-Bearing Quartz Veins Analyzed by Solid-Mass Spectrometry: Method and Contributions to the Orogenic Metallogenic Model and Exploration", by Damien Gaboury, focuses on a detailed study of the volatile composition of different fluid inclusions that occurred in the gold hydrothermal deposits from Canada, Burkina Faso, Sudan and New Zealand.

Chapter 4, "Solid Solution Thermochemistry: Understanding Metal Zoning in Polymetallic Hydrothermal Sulfide Ore Deposits", by Richard Sack, focuses on detailed thermochemical studies of fluid inclusions from different fahlores and sphalerite.

Chapter 5, "Alteration of Durbachites and U-Th Mineralization in the Durbachites of the Třebíč Pluton, Czech Republic", by Miloš René, focuses on mineralogy and geochemistry of small uranium deposits occurring in the eastern part of the Bohemian Massif, Czech Republic.

Miloš René
Institute of Rock Structure and Mechanics, v.v.i.,
Czech Academy of Sciences,
Prague, Czech Republic

Chapter 1

Introductory Chapter: Ore Geology – a Significant Part of Geosciences

Miloš René

1. Introduction

Ore geology is a branch of geosciences that deals with geological position, mineralogy and geochemistry of various ore deposits occurring on the Earth. The ore deposits occur in various igneous, metamorphic and sedimentary rock series. Ores are natural rocks that contain one or more valuable minerals concentrated above background levels, typically containing various metals that can be mined, treated and sold at a profit. The grade of ore refers to the concentration of the desired materials it contains. The presented book deals with ore deposits of gold, tantalum, niobium, polymetallic sulfides and uranium. The treatment of all these ore components is highly significant for our world community's recent technical and economic development.

2. Gold

The gold represents the earliest metal employed by humans. The oldest gold artifacts in the world are from Bulgaria, which date back to the fifth millennium BC. However, the first gold mining was documented in Ancient Egypt at the end of the fifth millennium BC. Large gold mines were also present across the Red Sea in the area of Saudi Arabia [1]. Gold, as a significant precious metal has been in human history, was used predominantly for coinage and jewellery. In the past, a gold standard was implemented in modern monetary policy. Gold coins were minted as a circulating currency in the 1930s; however, this world gold standard was abandoned for a fiat currency system in 1971.

In nature, gold often occurs in its native state as grains or nuggets in different magmatic and metamorphic rocks and hydrothermal veins in different alluvial ore deposits. In the hydrothermal ore deposits, gold also occurs in a solid solution together with silver as the electrum. In some of these hydrothermal ore deposits, gold also occurs in tellurides (e.g., calaverite, krennerite, nagyagite, petzite and sylvanite) [2].

3. Niobium and tantalum

Niobium and tantalum are two elements that have been used significantly in the recent industry. Niobium is primarily used in various alloys, with its most significant application being in the production of special steels for gas pipelines. The high-temperature stability of niobium-containing superalloys is important for its uses

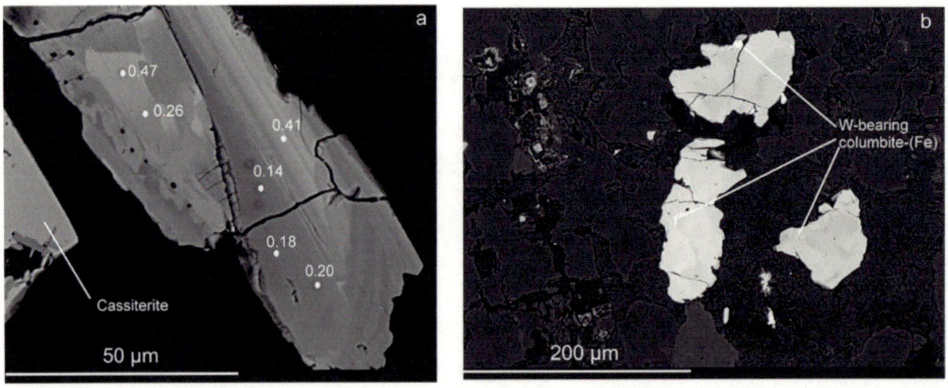

Figure 1.
BSE images of a zoned columbite-(Mn) showing variation in Ta/(Ta + Nb) ratio, b W-bearing columbite-(Fe) in Nb-Ta-bearing rutile [5].

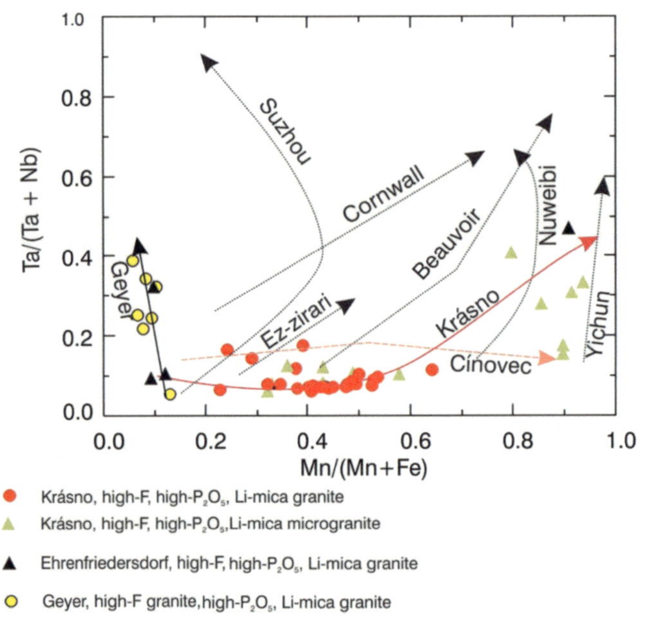

Figure 2.
Schematic Ta/(Ta + Nb) versus Mn/(Mn + Fe) composition plot for columbite-group minerals from high-F, high P_2O_5 and Li-mica granites from the Krušné hory/Erzgebirge Mts [6].

in jet and rocket engines. In nature, niobium occurs predominantly in the mineral columbite. However, niobium is also concentrated in the mineral pyrochlore ($(NaCa_2)Nb_2O_6(OH,F)$), which occurs in carbonatites. The world's leading producer of pyrochlore and also of niobium and ferroniobium is Brazil [3].

Recently, the most significant usages of tantalum are tantalum capacitors used in mobile phones and different electronic devices (computers and automobile electronics). In nature, tantalum is primarily concentrated in tantalite. The most significant world producer of tantalite is Democratic Republic of the Congo (about 35% of world production) [3]. In this case, tantalite occurs together with columbite, and the economically significant columbite-tantalite ore is known as coltan. This name is frequently used in connection with its mining in Africa, especially in Rwanda and Democratic Republic of the Congo (DRC). Notably, the mining of coltan in DRC is a significant matter for discussion of ethics. This is partly illegal and highly primitive mining [4].

The primary sources of columbite and tantalite in the above-mentioned coltan ores are hydrothermal ore deposits coupled with altered granitic rocks (greisens) and deposits of Sn-W ores. The nice examples of columbite contained in high-F granites, high P_2O_5 and Li-mica granites were published from the Krušné hory/Erzgebirge Mts (**Figures 1** and **2**) [5, 6].

4. Polymetallic sulfide ores

The polymetallic sulfide ores, as well as galenite, sphalerite and different sulfosalts, are also significant parts of very different scientific and economic studies. The detailed solid solution thermochemistry fahlore, presented in our book, is a very nice example of detailed mineralogical and geochemical studies.

5. Uranium

The Central European uranium deposits were the first industrially mined deposits in the world. The main uranium-bearing mineral, uraninite, was first used for the production of colours, which was used for the production of uranium-bearing glass and also for porcelain colouring. The silver metallurgical work in Jáchymov in the years 1851–1855 reconstructed the factory for the production of uranium yellow colours and also for producing uranium black colour, which was used in the porcelain industry. These uranium colours, produced in the Jáchymov factory, were highly valued on the European market. Therefore, in 1871, the factory was extensively reconstructed, and by 1879, the Jáchymov factory was recognized as the largest producer of uranium-based colours in the world. However, after the year 1896, the production of the uranium colours in this factory highly declined.

Uranium glass became popular in the mid-nineteenth century. The first major producer of uranium glass in this period was Franz Xaver Anton Riedel from the Dolní Polubný (Unter Polaun) in north Bohemia. This glassmaker started with the production of uranium glass in 1841 and developed yellow and green uranium glass. The highly different uranium glass products were presented predominantly as pressed glass cups and polished vases (**Figure 3**) [8, 9].

Figure 3.
Cup from uranium glass [7].

Otto Hahn and Fritz Strassman conducted experiments that led to the discovery of uranium's ability to fission in 1934. The discovery of nuclear fission led the United States, Great Britain and the Soviet Union to the development of nuclear weapons and nuclear power. These experiments started after World War II (1939–1945) with high exploration of uranium ore deposits. In Central Europe, the exploration of uranium ores and mining was concentrated in East Germany (GDR) and Czechoslovakia. In the GDR, the exploration and mining of uranium ores were concentrated in the Erzgebirge Mts., especially in the Schlema-Alberoda area. The total uranium production in 1946–1991 from this ore district formed by the vein U-Ag-Ni-Co-Bi ores was 80500.8 t of uranium (**Figure 4**) [11].

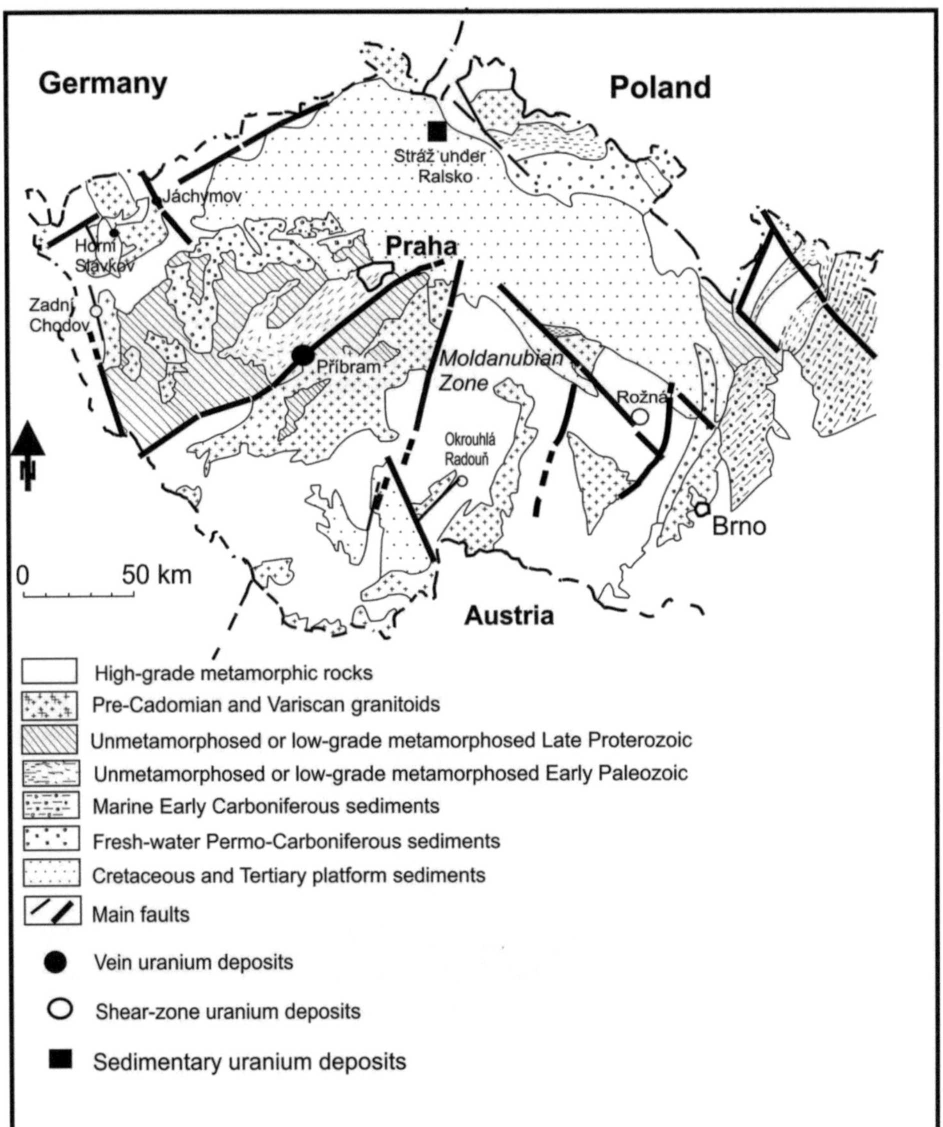

Figure 4.
Main uranium ore deposits in the Czech part of the Bohemian Massif [10].

In Czechoslovakia, the main concentrations of uranium ores were concentrated in the Bohemian Massif. The mined uranium deposits in this area could be divided into three genetic types: vein deposits, shear-zone deposits and sedimentary deposits. The vein deposits are predominantly represented by the Příbram, Jáchymov and Horní Slavkov deposits. The production of uranium from the biggest Příbram deposit mined in the years 1950–1991 was 50,200,8 t U. The shear-zone deposits are represented predominantly by the Rožná, Zadní Chodov and Okrouhlá Radouň ore deposits. The biggest Rožná uranium deposit was mined in the years 1957–2017 with a total production of 22,220 t U. The sedimentary uranium deposits are represented predominantly by the North Bohemian Cretaceous deposits in the area of the Stráž under Ralsko. This deposit was mined in the years 1967–1996 with a total production of 14,674,1 t U [12].

6. Conclusions

The introductory chapter discussed the geology of ore deposits, which are presented in detail in this book. This geology consists of gold, niobium, tantalum, sulfosalts, and uranium. Gold represents the earliest metal employed by humans, and its mining was very often coupled with variable human conflict. Niobium and tantalum have recently become highly significant parts of the world's critical minerals. The different hydrothermal sulfides are highly important matters of ore geology. The use of uranium was, in the past, a critical fuel for nuclear weapons and recently has significant importance for modern nuclear power.

Conflict of interest

There is no conflict of interest.

Author details

Miloš René
Institute of Rock Structure and Mechanics, v.v.i., Czech Academy of Sciences, Prague, Czech Republic

*Address all correspondence to: rene@irsm.cas.cz

References

[1] Sutherland CHV. Gold: Its Beauty, Power and Allure. London: Thames and Hudson; 1959. 196 p

[2] Spry P, Chryssoulis S, Ryan CG. Process mineralogy of gold: Gold from telluride-bearing ores. Journal of minerals, metals and materials society. 2004;**56**:60-62. DOI: 10.1007/s11837-004-0185-4

[3] U.S. Geological Survey. Mineral Commodity Summaries 2024. Reston, Virginia: U.S. Geological Survey; 2024. 212 p

[4] Democratic Republic of Congo. UN Panel on Plunder of Resources Publishes Final Report. New York: UN News; 2003. pp. 1-3

[5] René M, Škoda R. Nb-Ta-Ti oxides fractionation in rare-metal granites: Krásno-Horní Slavkov ore district, Czech Republic. Mineralogy and Petrology. 2011;**103**:37-48. DOI: 10.1007/s00710-011-0152-z

[6] René M. Nb-Ta-Ti oxides in topaz granites of the Geyersberg granite stock (Erzgebirge Mts., Germany). Minerals. 2019;**9**(155):1-14. DOI: 10.3390/min9030155

[7] A uranium glass beaker, Bohemia, c. 1840 [Internet]. 2024. Available from: www.dorotheum.com/en/l/8527735

[8] Kirchheimer F. Uranium and its History. Stuttgart: E. Schweizerbart'sche Verlagsbuchhandlung; 1963. 380 p. (In German)

[9] Schwanker RJ, Eigenstetter M, Laubinger R, Schmidt M. Uranium as color in glass and glaze. Physik in Unserer Zeit. 2005;**36**:100-167. (In German)

[10] René M. History of uranium mining in Central Europe. In: Uranium – Safety, Resources, Separation and Thermodynamic Calculations. London: IntechOpen; 2008. pp. 1-20

[11] Hiller A, Schuppan W. Geology and uranium mining in the district Schlema-Alberoda. In: Bergbau in Sachsen. Vol. 14. Freiberg: Sächsisches Landesamt für Umwelt und Geologie; 2008. 200 p

[12] Kafka J, editor. Czech Ore and Uranium Mining Industry. Ostrava: Anagram; 2003. 647 p. (In Czech)

Chapter 2

Geochemical Characterization of Cassiterite, Ferberite and Columbite-Tantalite from Rwinkwavu, Bugarura-Kuluti and Musha-Ntunga Mineral Districts of the South-Eastern Rwanda

Jean-Claude Ngaruye and Christoph Gauert

Abstract

The mines of Eastern Rwanda are historically known to host cassiterite, ferberite and niobo-tantalite mineralization. The lithology here is dominated by meta-sedimentary rocks of Mesoproterozoic age. The latter were deformed during the Kibaran Orogeny (1.4 to 1 Ga) and intruded by two generations of granite: G1–3 granites of ca. 1380±10 Ma and G4- granites of ca. 986±8 Ma. The economic minerals exploited seem to be likely related to the late phases of G4 granite venue which functioned as the heat source for the mineralizing fluids. SEM-EDS/WDS investigations on cassiterite samples revealed that the dominant substitutions were of the $Sn^{4+} \leftrightarrow (Ta, Nb)^{4+}$, 3 $Sn^{4+} \leftrightarrow 2(Ta, Nb)^{5+}$ +(Fe, Mn) $^{2+}$ and/or Fe^{3+} + $OH^{-} \leftrightarrow Sn^{4+}$ + O^{2-} types. Laboratory analysis of columbite-tantalite mineralization from increasingly distal veins showed evolution trends from ferro-columbite to mangano-tantalite compositions, indicating the increase of Ta and Mn with advanced differentiation of pegmatites. The fluid inclusion studies showed that the mineralizing fluids were composed of more than one type of cation. In addition, obtained salinities are below 17.5wt.% NaCl equiv. and T_h between 103 and 360°C. They confirmed that the columbite-tantalite precipitation closer to the granites was achieved in pneumatolytic conditions, whereas the cassiterite and ferberite mineralization precipitated in hydrothermal environments, possibly derived from a mixture between magmatic fluids and meteoric to connate waters. A conceptual metallogenetic model was developed and includes a multi-stage hydrothermal circulation induced by progressive granitic magmatism and precipitation of Sn, W and Nb-Ta minerals in pegmatitic veins cross-cutting lithologies acting as chemical traps.

Keywords: Kibaran Belt, south-eastern Rwanda, meta-sedimentary rocks, G4 granites, mineral chemistry, cassiterite, wolframite, columbite-tantalite, fluid inclusions, hydrothermal alteration, metallogenetic model

IntechOpen

1. Introduction

The study area covers three mining concessions located in the Eastern Province of the Republic of Rwanda, 45 km east of the capital city Kigali and in 20 km distance from the Akagera National Park. The three concerned mining concessions are recorded among the first producers of cassiterite and columbite/tantalite. They are Musha-Ntunga, Bugarura-Kuluti and Rwinkwavu (**Figure 1**).

The geological and mining research on the Kibaran belt have been carried out since the 1930s, but very few recent metallogenetic studies to explain the origin of the mineralization using modern techniques and metallogenetic models were reported [2–7]. The aim of this paper is to investigate the mineral chemistry of ore minerals and, using various analytical techniques, argue their petrogenetic significance and their importance as exploration tool for future mineral discoveries.

2. Geological setting

South-Eastern Rwanda is being taken by Precambrian rock suites that are part of the Mesoproterozoic Karagwe-Ankole Belt (KAB), spanning through Rwanda, Burundi, SW Uganda and NW Tanzania as well as the Kivu-Maniema region of the Democratic Republic of Congo (DRC) [8]. Tack et al. [8] report two structurally contrasting domains: the Western Domain (WD), hosting Rwanda and resting on a Paleoproterozoic basement and the Eastern Domain (ED), resting on an Archean basement. They are separated by a boundary zone, the Kabanga-Musongati (KM) alignment of mafic to ultramafic Bushveld-type layered complexes of approximately 1.4 Ga age.

The KAB shows a long-lived history of an intra-cratonic belt characterized by intermittent depositional activity with periods of hiatus of deposition, erosion, and magmatism.

The deposits indicate a recurrent subsidence trend controlled by structural activity from east to west. The KAB is the result of the approximately 1375 Ma regional "Kibaran tectono-magmatic event", the Kibara Large Igneous Province [9], due to

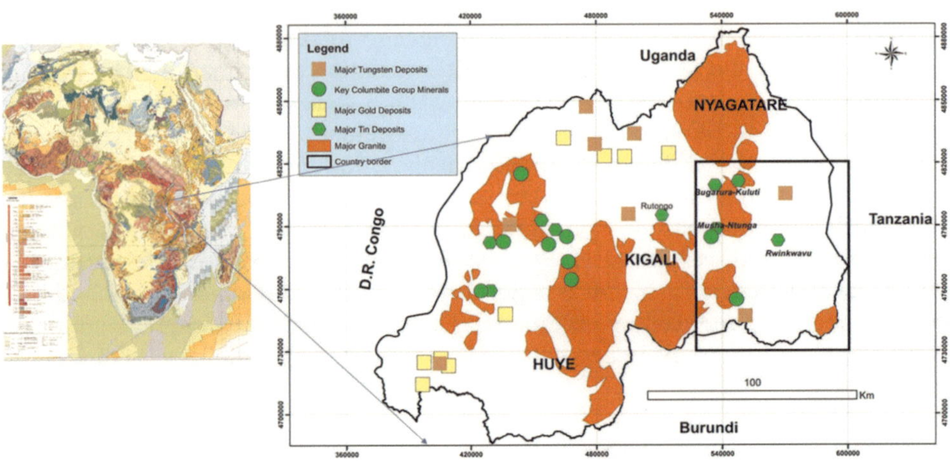

Figure 1.
Location map of the study area on the mineral deposits map of Rwanda (1:250000). The stars indicate the main mining concessions of the country [1].

intra-cratonic tectonic extension and opening of rift basins which started at about 1800 Ma before today. The long-lived history of the KAB as aulacogen of shallow-water intra-cratonic basin(s) within the proto-Congo craton was interrupted only twice by short-lived compressional deformation resulting from the effects of global orogenic events, outside of the proto-Congo craton [8]. The first compressional tectonic event (D1) is related to the formation of the Irumide & Chipata-Tete Belts during the Rodinian amalgamation at the transition between the Meso and Neoproterozoic (Stenian–Tonian). The second event (D2), which caused an northern to southern Pan-African overprint, took place during the late Neoproterozoic (Ediacaran) to early Cambrian and is related to the Gondwana amalgamation.

The deformation history of Rwanda is best known from the Gitarama-Gatumba axis in western Rwanda [10].

The degree of metamorphism extended from upper greenschist to lower amphibolite facies. Small folds striking between N20W and N60W were locally superimposed on the regional structure, mostly expressed by a crenulation foliation in the fine-grained sediments.

Two magmatic episodes are recognized in the area of study [8, 11, 12]:

- The first episode consists of an abundant 1380 to 1370 Ma old coeval bimodal magmatism emplaced under an extensional regime during the "Kibara tectono-magmatic event". This short-lived but prominent event has intra-plate characteristics, is rift-related and indicative of attempted, though unsuccessful, continental break-up. The felsic rock suite is present in the form of massive peraluminous S-type two-mica granitoids represented by the Rusumo and Lake Ihema granitoids (Tanzania border) in the vicinity of Rwinkwavu or Nyagatare batholiths of Bugarura-Kuluti area. They are historically known as "G1–G3 granites" and are devoid of economic minerals.

  The mafic rock suite is slightly represented in the South-Eastern part of the Rwinkwavu area and consists of limited outcrops of mafic to ultramafic intrusions [13].

- The second episode occurred during the Neoproterozoic (around 985 Ma) when S-type leucogranites (historically known as "G4 or tin-granites") were emplaced with accompanying Sn, Nb-Ta, W, and Au-bearing pegmatites and quartz veins which gave rise to the world-class Sn-metallogenic province overprinting large areas of the KAB [2, 6] hosting the South-Eastern Rwanda. Local contact metamorphism is observed around granite, pegmatite, and amphibolite bodies. These contact-metamorphic effects are often confined to the metasediments immediately adjacent to the magmatic intrusions,

There is tectonic, field, and geochronological evidence for a Paleoproterozoic (1.9–2.1 Ga) basement underlying the Akanyaru Supergroup. The basement is generally composed of ortho- and paragneisses of various compositions, metasediments and locally migmatites. These basement materials, together with Mesoproterozoic granites, are likely located in the cores of large antiform structures (**Figure 2**).

South-Eastern Rwanda is built up of sedimentary units belonging to the Akanyaru Supergroup which comprises three successive groups of, predominantly, siliciclastic composition and subordinate interlayered metavolcanics with a total thickness of between 5000 and 6000 m. They are, respectively, from bottom to top: the Gikoro,

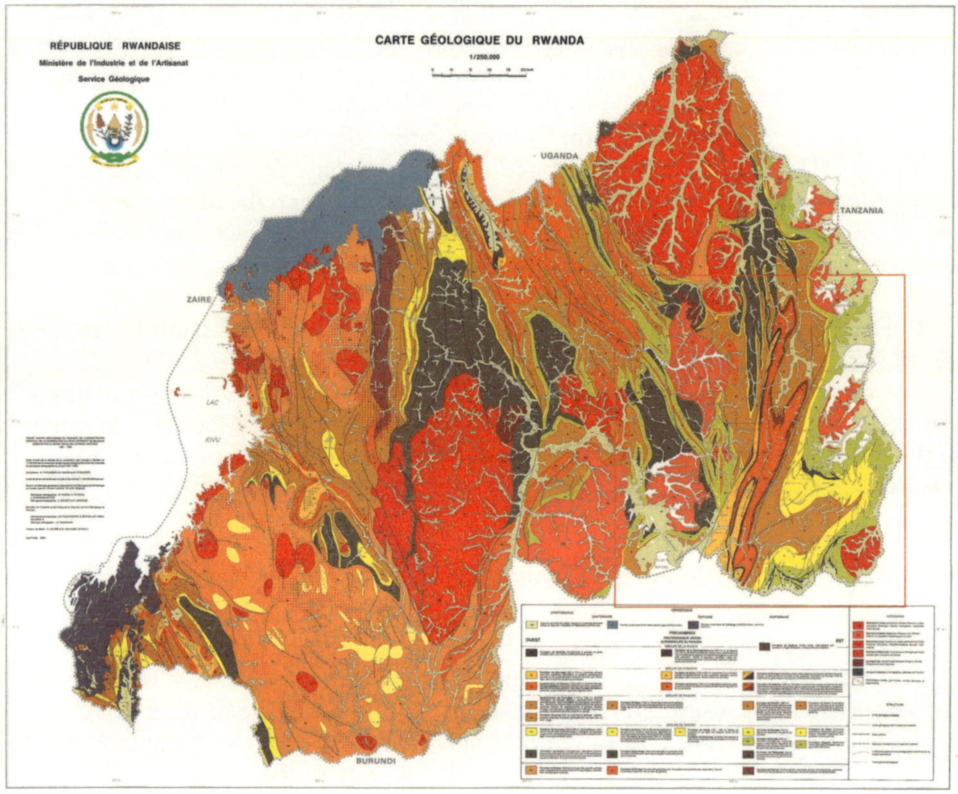

Figure 2.
Sketch Geological map of south-eastern Rwanda (modified after Theunissen et al. [14]).

Pindura and Cyohoha Groups. They consist of alternations of metasandstones, quartzites, phyllites and schists/shales (**Figure 3**). The depositional environments of the Akanyaru Supergroup indicate a shallow-water basin (or sub/basins) with deposition contexts ranging from mud-shelf (bottom) to sandy-shelf with possible emersion (top). The Gikoro and Pindura Groups were deposited between 1.42 Ga and 1.37 Ga (during the early Ectasian), and the Cyohoha Group was assumed to be deposited between 1.2 Ga and 1.0 Ga (largely during the Stenian period). The Rugezi Group, dominated by diamictites of Gisakura formation and outcropping in the south-west and north of the country, seems to be not present in the area of the current study.

Bimodal volcanic intercalations associated with the 1375 Ma event, best described in western Burundi, are present throughout the whole Akanyaru Supergroup with paroxysm in the Pindura Group where also carbonate lenses were formed during felsic volcanism.

In terms of mineral resources, the primary deposits of Rwinkwavu, Bugarura-Kuluti and Musha-Ntunga are genetically related to hydrothermal alteration in the form of pegmatite or quartz veins. The alluvial/eluvial tin deposits of the Kizanye-Rutonde trend are the most important and, generally, are composed of four layers of gravels/laterites separated by thin layers of clays. The artisan miners in Rutonde mining site confirmed that the cassiterite production per cubic meter ranges from 7 to 30 kg, and the lowest layers are the most mineralized. The Bugarura-Kuluti and Musha-Ntunga occurrences host also secondary sedimentary mineralizations.

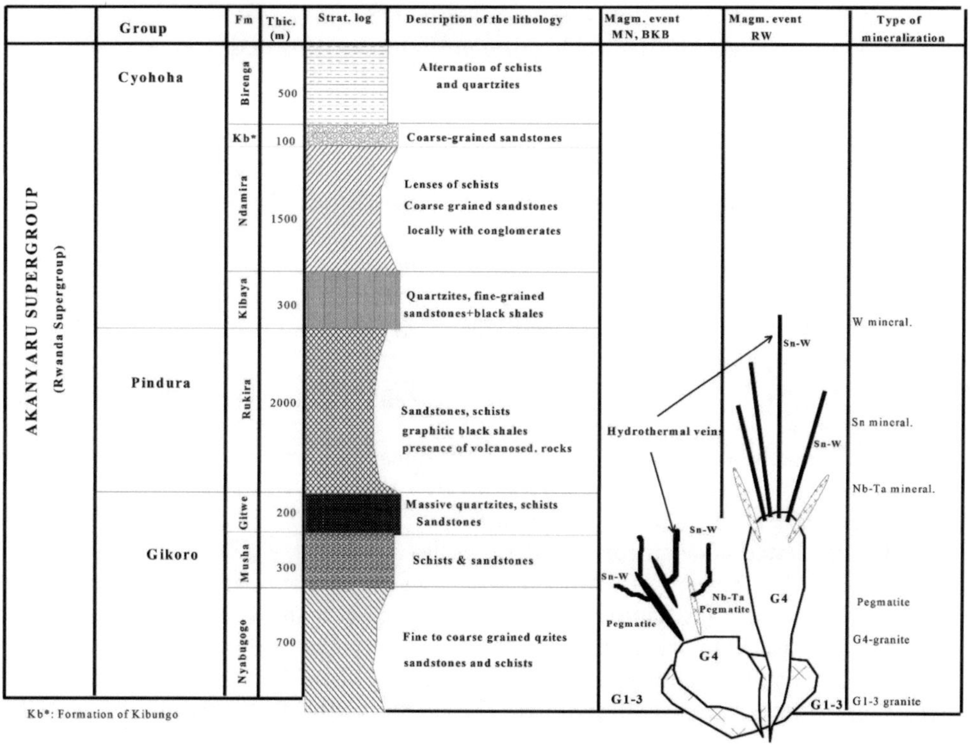

Figure 3.
Simplified stratigraphic column combined with magmatic and metallogenic events in the study area adapted after Dewaele et al. [15], Tack et al. [8] and Baudet et al. [16]. MN-Musha-Ntunga; BKB-Bugarura-Kuluti-Bibare; Fm-formation; Bur.-Burundi; Gat. -Gatumba.

3. Methodology

SEM-EDS/WDS: The samples of ore minerals were coated using a carbon coating machine, QUORUM/Q150T. The SEM-E/WDS analyses were completed on ore mineral concentrates and rock samples for the mineral chemistry determination with a SHIMADZU SSX-550 Superscan SEM of the Department of Geology and a JEOL JSM-6610 SEM provided by the Electron microscope unit of the Faculty of Science of the University of the Free State/Bloemfontein in South Africa (Dept. Geol., UFS, BFN, RSA) for this purpose. The SEM-EDS data were measured with internally calibrated standards and the results were automatically normalized to 100%. Approximately 200 qualitative and quantitative SEM-EDS analyses were performed on the 13 representative samples from Rwinkwavu, Musha-Ntunga and Bugarura-Kuluti mining areas.

Fluid inclusion studies: Doubly polished thick sections have been prepared and the transmitted light microscopy techniques using an OLYMPUS BX51 microscope were used for the first fluid inclusions mapping at room temperature (19–21°C). In addition to the microscopic investigations, microthermometric methods [17] were applied to the fluid inclusions above 10 μm using the LINKAM TMSG 600 stage made available by the Department of Geology, University of the Free State. Around one hundred freezing-heating measurements were made. Pure CO_2 inclusions in quartz were used for the calibration of the instrument at the triple point of CO_2 (−56.6°C). The selected fluid inclusions were first of all cooled down to between −70 and − 100°C at the rates of 10 to 20°C/minute. Then stepwise, they were heated up at the rates of 0.5 to 5°C/minute and the changes during the heating period were recorded.

4. Results

4.1 Mineral chemistry of cassiterite and wolframite from Rwinkwavu

4.1.1 Mineral chemistry of cassiterite from Rwinkwavu

Table 1 summarizes the results from SEM-EDS analyses of selected cassiterite samples in atoms per formula unit (apfu) which have been used in (Fe + Mn) versus (Ta + Nb) (**Figure 4**) and (Fe + Mn) versus W (**Figure 5**) discrimination diagram after Möller et al. [18].

In order to exclude various types of mainly siliceous impurities within the samples from various mineral occurrences, the ratio (Ta + Nb)/(Fe + Mn) was used for the discussion and this is illustrated in **Figure 4**.

Figure 4, plotted and interpreted after Möller et al. [18], shows the discrimination diagram of (Fe + Mn) versus (Ta + Nb) for the selected investigated cassiterite samples from Rwinkwavu expressed in apfu. It shows a poor correlation between Ta + Nb and Fe + Mn. If the element contents are assumed to be due to contamination by mineral inclusions of the type (Fe, Mn) $(Ta, Nb)_2O_6$, then, a correlation along the line which indicates the atomic ratio (Ta + Nb)/(Fe + Mn) = 2 should be expected.

However, the majority of investigated analysis fall below a line that indicates an atomic ratio (Ta + Nb)/(Fe + Mn) = 2. There is, thus, a clear excess of (Fe + Mn) content which could indicate the presence in cassiterite crystals of an Fe-rich inclusion testifying to the possible substitution between Sn and Fe based on the $Sn^{4+} \leftrightarrow Fe^{3+} + H^{+}$ or $Sn^{4+} + O^{2-} \leftrightarrow Fe^{3+} + OH^{-}$ substitution law.

The remaining three analyses are plotting along the line (Ta + Nb)/(Fe + Mn) = 2 indicating that (Ta + Nb) content in cassiterite is controlled by the solid inclusions of the type (Fe, Mn) $(Ta, Nb)_2O_6$ and therefore, follows the substitution law of the type $3\,Sn^{4+} \leftrightarrow 2\,(Ta, Nb)^{5+} + (Fe, Mn)^{2+}$.

The above correlation diagram shows a poor positive correlation of (Fe + Mn) with W. The field survey did not show any outcrop of felsic igneous rocks generating the rare element ores such as cassiterite and wolframite exploited in Rwinkwavu. Therefore, all the cassiterite samples were hand-picked from either muscovite-sericite-quartz vein originated from B-rich hydrothermal fluids (tourmalinization of the host meta-sedimentary rocks on 20 cm to 40 cm) or from placers and thus, may represent the cassiterites crystallized under hydrothermal conditions. **Figures 4** and **5** were adapted from Möller et al. [18] and reflect the data from **Table 1** where the values of W/Sn and (Fe + Mn) taken into account are those which are above zero. Apart from the analyzed points of the only cassiterite sample from Gahengeri, no. JC-23-6-6638, which is plotted along the line of atomic ratio W/(Fe + Mn) = 1 and thus, indicating the presence of wolframite in solid inclusions, the all-remaining samples fall above the diagonal line of eq. W/(Fe + Mn) = 1. Contrary to the previous correlation diagram (**Figure 5**) W content occurs in evident excess instead of Fe + Mn and consequently, (Fe + Mn) seems to be independent from contamination of the type (Fe, Mn) WO_4. This might indicate that the substitution of the type $Sn^{4+} \leftrightarrow W^{4+}$ predominates in the cassiterite samples from Rwinkwavu.

Sample	JC-GHR					JC-NYZ					JC-38						JC-455
Analysis	2	4	5	15	19	2	4	5	6	7	2	3	5	6	7	8	5
Nb_2O_5	0.1	0.0	0.6	0.1	0.0	0.0	0.0	0.0	0.0	0.0	0.1	0.4	0.0	0.2	0.5	0.0	0.3
Ta_2O_5	0.0	0.7	0.2	0.0	0.6	0.7	0.9	1.0	1.1	1.1	1.1	0.0	0.3	0.5	0.0	0.7	0.3
SnO_2	97.5	96.9	98.9	99.1	96.7	82.8	90.3	90.2	91.2	80.7	98.2	98.5	99.3	98.6	98.4	98.6	98.9
SiO_2	0.0	0.0	0.0	0.0	0.0	6.1	0.0	0.0	0.0	4.5	0.0	0.0	0.0	0.0	0.0	0.0	0.0
TiO_2	0.0	0.0	0.2	0.1	1.1	0.0	0.4	0.0	0.0	0.3	0.4	0.6	0.1	0.2	0.2	0.3	0.2
Fe_2O_3	0.7	0.7	0.2	0.7	0.6	1.1	0.5	1.1	1.1	0.7	0.2	0.5	0.2	0.3	0.2	0.2	0.2
Al_2O_3	1.5	1.7	0.0	0.0	1.0	2.9	1.6	0.8	0.5	0.4	0.0	0.0	0.0	0.0	0.7	0.0	0.0
F	0.0	0.0	0.0	0.0	0.0	0.0	0.0	0.0	0.0	6.3	0.0	0.0	0.0	0.0	0.0	0.0	0.0
Tot.	100.8	100.7	100.8	100.8	100.8	100.4	100.4	100.5	100.5	100.3	100.5	100.5	100.5	100.5	100.5	100.5	100.0
Number of cations based on two oxygens																	
Nb	0.001	0.00	0.006	0.001	0.000	0.00	0.00	0.00	0.00	0.00	0.001	0.004	0.000	0.003	0.006	0.000	0.004
Ta	0.000	0.005	0.00	0.000	0.004	0.005	0.006	0.007	0.007	0.007	0.007	0.000	0.002	0.003	0.000	0.005	0.002
Sn	0.967	0.960	0.98	0.982	0.959	0.821	0.895	0.894	0.904	0.799	0.973	0.977	0.984	0.977	0.975	0.977	0.980
Si	0.00	0.001	0.00	0.001	0.001	0.152	0.000	0.000	0.000	0.112	0.000	0.000	0.000	0.000	0.000	0.000	0.000
Ti	0.001	0.001	0.0	0.001	0.020	0.000	0.008	0.000	0.000	0.005	0.008	0.011	0.001	0.003	0.004	0.006	0.004
Fe	0.014	0.012	0.0	0.013	0.01	0.021	0.009	0.020	0.020	0.012	0.003	0.008	0.004	0.006	0.003	0.004	0.004
Al	0.044	0.050	0.0	0.001	0.030	0.085	0.045	0.024	0.013	0.010	0.001	0.001	0.001	0.001	0.020	0.001	0.000
ΣCat.	1.038	1.039	1.006	1.010	1.034	1.155	1.008	1.006	1.000	0.988	1.000	1.008	1.001	1.003	1.015	1.001	0.995
Nb/Ta	4.988	0.074	6.319	4.988	0.109	0.047	0.039	0.033	0.031	0.030	0.151	33.256	0.104	0.831	44.895	0.045	1.824
Nb + Ta	0.002	0.005	0.007	0.002	0.005	0.005	0.006	0.007	0.008	0.008	0.009	0.005	0.002	0.006	0.006	0.005	0.006
Fe + Mn	0.015	0.014	0.004	0.014	0.012	0.021	0.010	0.021	0.020	0.013	0.003	0.009	0.006	0.010	0.003	0.004	0.006
(Nb + Ta)/Sn	0.00	0.01	0.01	0.00	0.00	0.01	0.01	0.01	0.01	0.01	0.01	0.00	0.00	0.01	0.01	0.01	0.01

Sample	JC-GHR					JC-NYZ					JC-38						JC-455
Analysis	**2**	**4**	**5**	**15**	**19**	**2**	**4**	**5**	**6**	**7**	**2**	**3**	**5**	**6**	**7**	**8**	**5**
(Fe + Mn)/Sn	0.02	0.01	0.0	0.01	0.01	0.03	0.01	0.02	0.02	0.02	0.00	0.01	0.01	0.01	0.00	0.00	0.01
W/Sn	0.00	0.00	0.0	0.00	0.00	0.00	0.00	0.00	0.00	0.00	0.00	0.00	0.00	0.00	0.00	0.00	0.00

Table 1.
Chemical composition of selected cassiterite samples from Rwinkwavu.

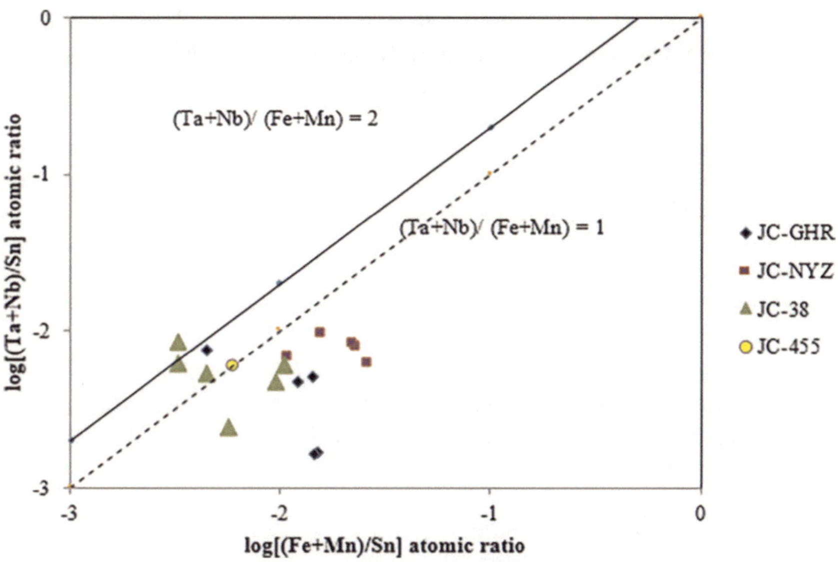

Figure 4.
Correlation diagram of cassiterite samples of Rwinkwavu using log [(Ta + Nb)/Sn] versus log [(Fe + Mn)/Sn] after Möller et al. [18].

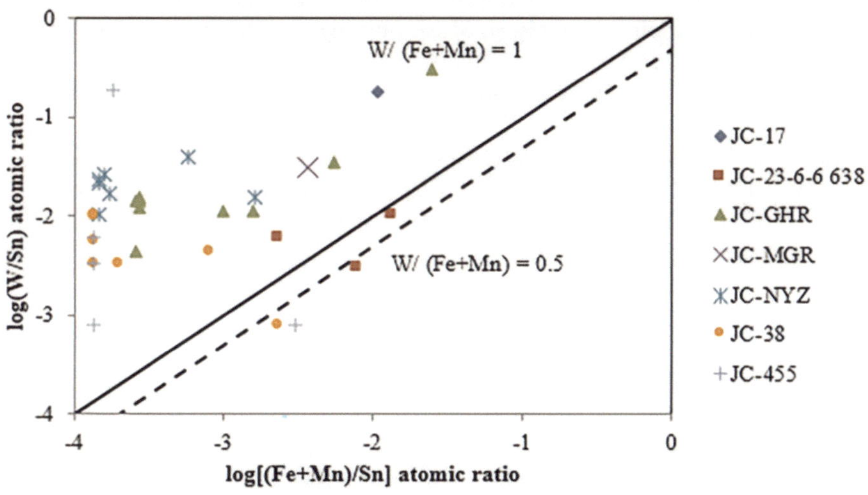

Figure 5.
Correlation diagram for selected cassiterite samples using log [W/Sn] versus log [(Fe + Mn)/Sn] after Möller et al. [18].

4.1.2 Mineral chemistry of wolframite in Rwinkwavu surroundings

No representative wolframite samples could be obtained from mining operations in the Nyamuyorwa tungsten ore deposit due to mine closure because of technical problems.

However, two wolframite samples from a primary quartz vein in the surroundings of Nyamuyorwa were hand-picked and analyzed. It is assumed that the chemical and mineralogical characteristics of both wolframite mineral deposits (Nyamuyorwa and

Kirimbari) may show similarities, firstly based on the similar geological contexts of the quartz vein type ore deposits, the location in the same Rukira formation and within the same Kirimbari-Rwinkwavu anticlinal structure and secondly, because of the absence of outcrop of granite parental to rare metals.

Additional to these three element oxides, small amounts of Br, P_2O_5, Al_2O_3, Ta_2O_5, SnO_2, CaO and MgO were identified and quantified in some of the selected analysis of the investigated wolframite samples (**Tables 2–4**).

The formula for wolframite is (Mn, Fe)WO_4. For a better chemical characterization of wolframite grains of two samples from the Kirimbari-Rwinkwavu anticline, a discrimination diagram of Fe-ferberite (in apfu) against Mn-huebnerite (in apfu) in **Figure 6** displays the Fe-dominance of the solid solution.

This formula will be correct in the case of the distribution of the analyzed analysis along the line which corresponds to Mn/Fe = 1. Contrarily, all the analyses are plotting far below the Mn/Fe = 1 line and most of them plot below the line Mn/Fe = 0.25, indicating that Fe- end-members are highly dominant and, therefore, indicates that the analyzed wolframite is a ferberite.

A negative trend exists, which points towards continuous substitution between the Fe and Mn end-members of this solid solution series.

4.2 Mineral chemistry of cassiterite and columbite-tantalite from Musha-Ntunga

In order to examine the chemical and mineralogical variations in the Sn- and Nb-Ta-bearing minerals and their host rocks in Musha-Ntunga mining concession, seven (7) samples were prepared and analyzed using the SEM/EDS-WDS techniques. These include two samples of tin mineralization and one sample of columbite-tantalite. A total of around 80 mineral grains were analyzed and the results are displayed in **Tables 5** and **6**. Additional to the cassiterite and columbite-tantalite minerals, the SEM/EDS-WDS investigations revealed that the mineral associations in those samples comprise also inclusions of quartz, biotite, muscovite, garnet, tourmaline, Fe-minerals, rutile, ilmenite, zircon and sulfides (chalcopyrite, arsenopyrite and pyrite). This section will focus specifically on the mineral chemistry of the economically exploited ore minerals found in Musha-Ntunga: Cassiterite and columbite-tantalite.

4.2.1 Mineral chemistry of cassiterite

The cassiterite group minerals of Musha-Ntunga appear to be mainly composed of tin oxide (average of 96.95 wt. %), iron (0.87 wt. % of Fe_2O_3), 0.81 wt. % Al_2O_3 and minor amounts of MnO (0.1 wt. %), Nb_2O_5 (0.2 wt. %).

The majority of investigated mineral grains are identified as almost pure tin oxide minerals with little amounts of trace elements or inclusions of other mineral phases.

Furthermore, two analyses of sample JC-22-6-6639 (7 and 10) were Fe-rich cassiterite where the wt. % of total Fe as Fe_2O_3 is respectively 29.6 and 37.2 (without crystal water measurements) and corresponds possibly to the hydro-cassiterite (Sn, Fe) (O, $OH)_2$ as defined by Rösler [19]. Bivariate diagrams of Sn-versus Nb-, Ta-, Fe- and Ti- (**Figures 7–10**) of selected cassiterite samples from Musha-Ntunga showed poor correlations, slightly negative for Ta, Fe and Ti concentrations versus Sn content (**Figures 8–10**).

With the aim to understand a systematic chemical variation of analyzed cassiterite samples, the results were plotted in variation diagrams of log [(Ta + Nb)/Sn] vs. log [(Fe + Mn)/Sn] and log [W/Sn] versus log [(Fe + Mn)/Sn] after Möller et al. [18]

DOI: http://dx.doi.org/10.5772/intechopen.1008342

Analysis	5	6	7	8	9	10	11	12	13	16	17	18	20	21	23	28
WO_3	74.1	73.9	74.9	74.8	68.6	67.6	68.5	75.5	68.8	76.7	68.2	58.9	71.6	41.1	73.7	73.2
MnO	5.3	1.1	3.8	4.7	—	2.8	0.4	3.1	0.2	3.0	2.4	—	5.3	0.9	3.6	1.4
Fe_2O_3	19.9	24.6	21.1	20.1	30.2	28.9	29.6	21.3	29.7	20.3	28.9	36.5	20.6	56.8	22.2	25.4
CaO	0.2	—*	—	—	—	0.3	—	0.1	0.1	—	0.2	—	0.2	—	0.2	—
MgO	0.3	0.3	0.2	0.1	0.2	0.3	0.0	—	0.3	—	0.2	—	0.4	—	0.3	—
Ta_2O_5	—	—	0.0	0.3	1.0	—	1.2	—	0.9	—	0.2	—	1.9	0.3	—	—
Number of cations based on four oxygens																
W	0.900	0.898	0.909	0.909	0.833	0.821	0.832	0.917	0.835	0.931	0.829	0.716	0.870	0.499	0.895	0.889
Mn	0.209	0.046	0.152	0.187	—	0.111	0.015	0.123	0.010	0.119	0.094	—	0.211	0.037	0.141	0.055
Fe	0.702	0.867	0.743	0.707	1.064	1.019	1.044	0.749	1.047	0.716	1.017	1.287	0.725	2.001	0.782	0.895
Ca	0.009	—	—	—	—	0.016	—	0.007	0.005	—	0.008	—	0.009	—	0.012	—
Mg	0.021	0.020	0.013	0.007	0.012	0.022	0.002	—	0.019	—	0.013	—	0.030	—	0.024	—
Ta	—	—	0.000	0.003	0.013	—	0.016	—	0.012	—	0.002	—	0.024	0.004	—	—
$\sum$Cat.	1.71	1.71	1.7	1.7	1.79	1.853	1.782	1.692	1.787	1.664	1.835	2.12	1.71	2.39	1.72	1.73

Not detected.

Table 2.
Chemical composition of selected wolframite samples from Rwinkwavu – Kirimbari (Sample JC-KRM) using SEM-EDS/WDS.

Analysis	W	Mn	Fe	Ca	Mg	Ta	Sn	Al	Bulk chemical formulae
5	0.90024	0.20913	0.70189	0.00906	0.02113	—	0.00355	—	$Mn_{0.21}Fe_{0.7}W_{0.9}O_4$
6	0.89782	0.04563	0.86742	—	0.02043	—	—	—	$Mn_{0.04}Fe_{0.87}W_{0.9}O_4$
7	0.90935	0.15159	0.74310	—	0.01338	0.00013	—	—	$Mn_{0.15}Fe_{0.74}W_{0.9}O_4$
8	0.90874	0.18730	0.70682	—	0.00704	0.00344	—	—	$Mn_{0.2}Fe_{0.7}W_{0.9}O_4$
9	0.83309	—	1.06429	—	0.01197	0.01275	—	—	$Mg_{0.01}Ta_{0.01}Fe_{1.1}W_{0.8}O_4$
10	0.82143	0.11071	1.01850	0.01610	0.02184	—	—	—	$Ca_{0.02}Mn_{0.11}Fe_1W_{0.8}O_4$
11	0.83224	0.01508	1.04421	—	0.00211	0.01555	0.00336	—	$Ta_{0.01}Mn_{0.01}Fe_1W_{0.8}O_4$
12	0.91664	0.12301	0.74944	0.00654	—	—	0.00019	—	$Mn_{0.12}Fe_{0.75}W_{0.9}O_4$
13	0.83515	0.00952	1.04668	0.00503	0.01902	0.01160	—	—	$Ca_{0.005}Mg_{0.02}Ta_{0.01}Mn_{0.01}FeW_{0.8}O_4$
16	0.93109	0.11865	0.71633	—	—	—	—	—	$Mn_{0.12}Fe_{0.7}W_{0.93}O_4$
17	0.82884	0.09365	1.01709	0.00805	0.01268	0.00217	—	—	$Mn_{0.1}Fe_1W_{0.8}O_4$
18	0.71577	—	1.28722	—	—	—	—	0.24915	$Al_{0.25}Fe_{1.29}W_{0.72}O_4$
20	0.86976	0.21111	0.72549	0.00906	0.02958	0.02384	—	—	$Mn_{0.21}Fe_{0.7}W_{0.87}O_4$
21	0.49888	0.03730	2.00144	—	—	0.00370	—	—	$Mn_{0.04}Fe_2W_{0.5}O_4$
23	0.89454	0.14127	0.78184	0.01207	0.02395		—	—	$Mn_{0.14}Fe_{0.8}W_{0.9}O_4$
28	0.88907	0.05476	0.89524	—	—	—	—	—	$Ca_{0.01}Mg_{0.02}Mn_{0.05}Fe_{0.9}W_{0.8}O_4$
	0.84267	0.09429	0.96044	0.00412	0.01145	0.00457	0.00044	0.01557	Avg bulk formulae for wolfr.*

**Bulk chemical formula of the tungsten ore in Kirimbari (JC-KRM): $Al_{0.015}Mg_{0.114}Ca_{0.004}Mn_{0.094}Fe_{0.960}W_{0.843}O_4$.*

Table 3.
Determination of the bulk chemical formulae of selected wolframite samples from Kirimbari (JC-KRM and JC-KRM-1676): JC-KRM.

	14	15	16	17	19	21	22	23	26	27	28	29	34
CaO	1.03	0.33	5	0.44	—	0.29	—	—	—	—	—	—	—
MnO	5.24	3.55	—	3.88	2.32	3.14	2.01	—	2.07	—	0.99	0.79	2.95
Fe_2O_3	26.56	26.5	38.07	26.82	33.64	31.66	28.9	31.29	28.34	59.88	42.22	25.57	26.9
WO_3	64.62	68.71	56.93	66.21	62.82	63.14	69.09	68.71	68.62	39.08	52.88	72.56	68.75
P_2O_5	2.18	0.92	—	1.57	1.22	1.05	—	—	0.97	—	1.03	1.08	1.41
Number of cations based on four oxygens													
Ca	0.049	0.016	0.237	0.021	—	0.014	—	—	—	—	—	—	—
Mn	0.196	0.133	—	0.145	0.087	0.117	0.075	—	0.077	—	0.037	0.03	0.11
Fe	0.881	0.879	1.263	0.89	1.116	1.05	0.959	1.038	0.94	1.987	1.401	0.848	0.893
W	0.739	0.786	0.651	0.757	0.719	0.722	0.79	0.786	0.785	0.447	0.605	0.83	0.787
P	0.081	0.034	—	0.059	0.046	0.039	—	—	0.036	—	0.039	0.04	0.053
∑Cat.	1.955	1.848	2.151	1.944	1.967	1.961	1.824	1.824	1.839	2.488	2.14	1.748	1.842

Table 4.
JC-KRM-1676.

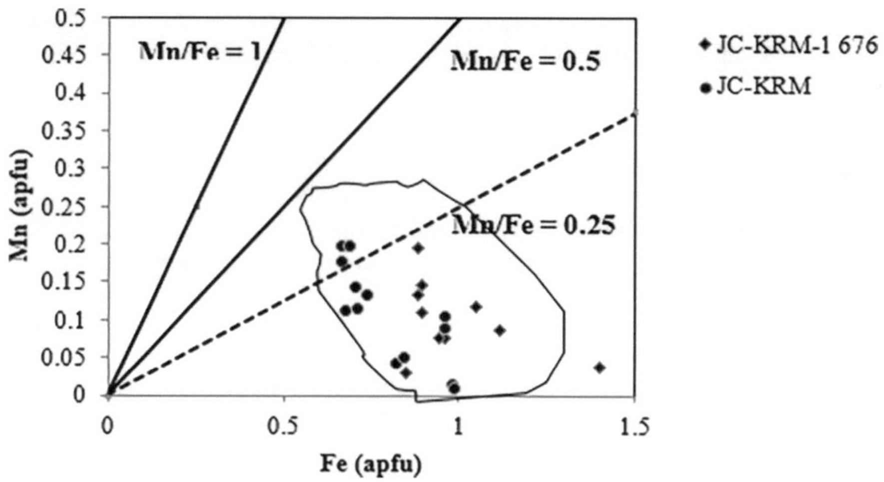

Figure 6.
Binary plot of Fe versus Mn (in apfu) for wolframite data from the study area.

(**Figures 11** and **12**). According to them, two types of compositions are distinguishable: 4 analyses are distributed along the line (Ta + Nb)/(Fe + Mn) = 2 and, therefore, reflecting contamination by mineral inclusions of the type (Fe, Mn) $(Ta, Nb)_2O_6$ and 3 points falling below that line. This latter case presents an excess of (Fe + Mn) content and indicates most likely the presence of Fe-rich mineral phases in the cassiterite.

Figure 12 shows that the two analyses of the cassiterite samples from Musha-Ntunga plot above the line of eq. W/(Fe + Mn) = 1 and therefore enriched in W, indicating that the relationship between Sn and W in the cassiterite crystal is mostly guided by a substitution of the type $Sn^{4+} \leftrightarrow W^{4+}$.

	JC-22-6-6639	JC-535	JC-25-6-3639	Average	LLD
SnO_2	97.4	95.19	98.27	96.95	1.95
Fe_2O_3	0.3	2.32	<LLD	0.87	0.25
TiO_2	0.5	<LLD	<LLD	<LLD	0.28
MnO	<LLD	<LLD	0.13	0.10	0.10
Nb_2O_5	<LLD	0.32	0.11	0.20	0.11
Ta_2O_5	0.6	<LLD	<LLD	<LLD	0.35
WO_3	0.4	<LLD	ND	<LLD	0.20
ZrO_4	0.1	0.10	<LLD	0.09	0.08
Al_2O_3	0.3	1.04	1.11	0.81	0.06
F	0.3	0.46	ND	0.24	0.09

Abbreviations: <LLD: Below the lower limit of detection; ND-Not detected; LLD-Lower limit of detection.

Table 5.
Average chemical compositions of cassiterite group minerals from Musha - Ntunga (JEOL JSM-6610 SEM-EDS in wt. %).

The remaining analyses of these samples fall below W/(Fe + Mn) = 0.5, and according to Möller et al. [18], this ratio suggests a high amount of Fe_2WO_6 in solid solution.

4.2.2 Mineral chemistry of columbite-tantalite

A systematic study of the columbite-tantalite mineralization from Musha-Ntunga was carried out on a thick section. Investigated areas of the sample were photographed by back-scattered electron image (BSE), structural formulae were calculated and the results were plotted (per analysis) in the "columbite quadrilateral" after Beurlen et al. [20]. All the analytical data are displayed in **Table 7**.

The results of the analyses (**Table 10**) show that the "columbite-tantalite" mineral is mostly composed of tantalum, averaging 41.10±2.5 wt. % Ta_2O_5, niobium, with 36.50 ±0.9 wt. % Nb_2O_5 in average, Fe_2O_3 with 9.93±0.7 wt. % and manganese with averages of 7.90±0.66 wt. % of MnO.

The bulk structural formulae and the main group in which those analyzed Nb- and Ta- bearing minerals are found are compiled in **Table 8**.

The averages of Nb/Ta and Fe/Mn ratios of the columbite-tantalite sample are less than 1 which means enrichment in both Ta and Mn. In addition, the Nb/Ta ratio less than 1 allows classification of this sample in the tantalite group. Based on Beurlen et al. [20], the mineral chemical analyses are reported in the four-pole quadrilateral diagram representing the end-member species in **Figure 13**. The columbite-tantalite sample from Musha-Ntunga shows a Ta/(Ta + Nb) ratio varying between 0.243 and 1 and Mn/(Mn + Fe) ratio ranging between 0.0 and 0.78. In the Beurlen et al. [20] diagram, there is a clear negative correlation between the Ta-Fe- and Nb-Mn- end-members and the analyses are plotting between the Ta- (tantalite group), Fe-(ferro-tantalite and ferro-columbite groups) and Mn-(Mangano-columbite and mangano-tantalite groups) rich end-members.

Sample			**JC-22-6-6639**					**JC-535**				**JC-25-6-3639**		
Point	**1**	**2**	**3**	**4**	**5**	**9**	**1**	**4**	**8**	**9**	**10**	**3**	**5**	**6**	**7**
WO_3	—	0.2	0.9	0.1	1.2	—	0.1	0.4	—	—	—	—	—	—	—
Nb_2O_5	0.5	—	—	0.3	—	—	0.6	—	0.8	0.3	—	—	—	—	0.5
Ta_2O_5	1.4	0.4	0.7	0.9	—	—	—	—	1.0	—	—	—	—	—	0.8
SiO_2	—	—	—	—	—	—	—	—	—	—	—	—	—	—	—
TiO_2	0.3	0.5	0.7	0.3	0.6	0.5	0.1	0.0	0.1	0.3	0.1	0.3	—	—	0.2
ZrO_2	—	—	0.2	0.2	—	0.4	—	—	—	0.5		0.2	—	—	—
SnO_2	95.9	97.8	96.1	97.9	98.0	98.6	99.1	99.6	95.9	86.2	95.2	98.3	100.0	98.3	96.4
Al_2O_3	—	0.8	0.9	—	—	—	—	—	—	2.8	2.4	1.1	—	1.7	1.7
Fe_2O_3	0.2	0.4	0.3	0.2	0.1	0.5	—	0.0	—	9.3	2.2	—	—	—	0.1
MnO	0.1	—	0.4	—	0.1	0.1	0.1	0.0	—	0.2	0.1	0.2	—	—	0.4
Tot	100.0	100.0	100.0	100.0	100.0	100.0	100.0	100.0	100.0	100.0	100.0	100.0	100.0	100.0	100.0
Number of cations based on two oxygens															
W	0.000	0.001	0.005	0.001	0.008	0.000	0.001	0.002	0.000	0.000	0.000	0.000	0.000	0.000	0.000
Nb	0.005	0.000	0.000	0.003	0.000	0.000	0.006	0.000	0.008	0.003	0.000	0.000	0.000	0.000	0.005
Ta	0.009	0.003	0.005	0.006	0.000	0.000	0.000	0.000	0.007	0.000	0.000	0.000	0.000	0.000	0.005
Si	0.000	0.000	0.000	0.000	0.000	0.000	0.000	0.000	0.000	0.000	0.000	0.000	0.000	0.000	0.000
Ti	0.005	0.009	0.012	0.006	0.011	0.008	0.002	0.000	0.001	0.005	0.001	0.005	0.000	0.000	0.004
Sn	0.929	0.947	0.931	0.948	0.949	0.955	0.960	0.964	0.929	0.835	0.922	0.952	0.969	0.952	0.934
Al	0.000	0.022	0.026	0.000	0.000	0.000	0.000	0.000	0.000	0.080	0.069	0.030	0.000	0.049	0.048
Fe	0.004	0.007	0.005	0.004	0.002	0.008	0.000	0.001	0.000	0.170	0.041	0.000	0.000	0.000	0.001
Mn	0.001	0.000	0.007	0.000	0.003	0.001	0.002	0.000	0.000	0.004	0.001	0.003	0.000	0.000	0.007
Σcat.	0.955	0.988	0.993	0.971	0.972	0.978	0.971	0.968	0.945	1.111	1.034	0.993	0.969	1.001	1.005

Sample	JC-22-6-6639						JC-535					JC-25-6-3639			
Point	**1**	**2**	**3**	**4**	**5**	**9**	**1**	**4**	**8**	**9**	**10**	**3**	**5**	**6**	**7**
Nb/Ta	0.570	0.000	0.000	0.548	—	—	—	—	1.260	—	—	—	—	—	0.935
(Nb + Ta)/Sn*	0.016	0.003	0.005	0.009	0.000	0.000	0.006	0.000	0.016	0.004	0.000	0.000	0.000	0.000	0.011
Log (*)	−1.80	−2.533	−2.286	−2.02	—	—	−2.193	—	−1.798	−2.418	—	—	—	—	−1.961
(Fe + Mn)/Sn**	0.006	0.008	0.013	0.004	0.005	0.010	0.002	0.001	0.000	0.209	0.045	0.003	0.000	0.000	0.009
Log (**)	−2.24	−2.124	−1.898	−2.35	−2.34	−1.98	−2.669	−2.927	—	−0.681	−1.34	−2.49	—	—	−2.052
W/Sn***	0.000	0.001	0.006	0.001	0.008	0.000	0.001	0.002	0.000	0.000	0.000	0.000	0.000	0.000	0.000
Log (***)	—	−2.899	−2.235	−3.06	−2.09	—	−3.069	−2.629	—	—	—	—	—	—	—

*Abbreviations: *(Nb + Ta)/Sn; **(Fe + Mn)/Sn; ***W/Sn; and —: not determined.*

Table 6.
Chemical composition of selected cassiterite samples from the Musha-Ntunga.

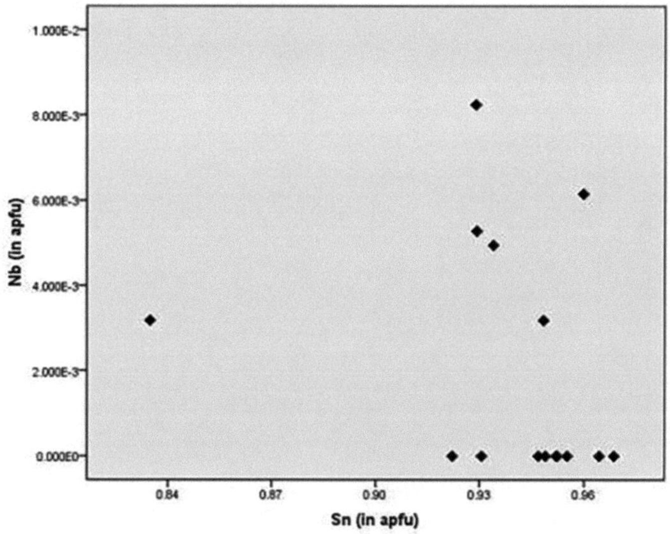

Figure 7.
Discrimination plot of Sn versus Nb (in apfu) in cassiterite of Musha-Ntunga.

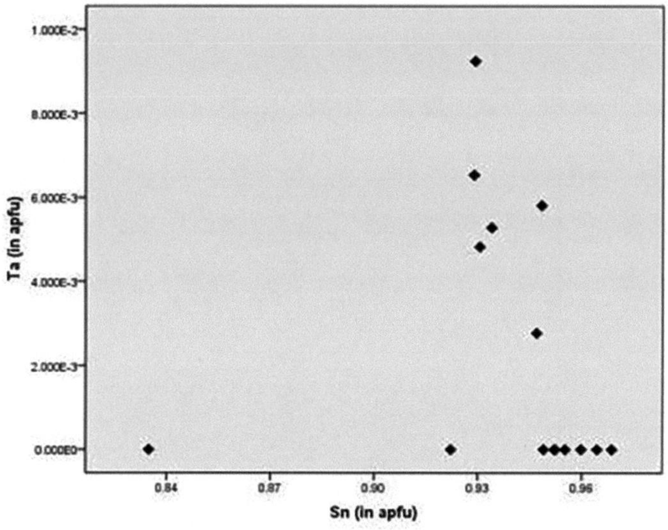

Figure 8.
Discrimination plot of Sn versus Ta (in apfu) in cassiterite of Musha-Ntunga.

4.3 Bugarura-Kuluti: Mineral chemistry of cassiterite, columbite-tantalite and wolframite

Chemical and mineralogical variations in the Sn-, W- and Nb-Ta- bearing minerals in the Bugarura-Kuluti mining district were discussed based on the results from SEM-EDS/WDS analyses performed on six (6) ore mineral samples. These include JC-36 from Rwankuba tungsten mine, JC-34 with Sn-Nb-Ta and four Sn-bearing samples.

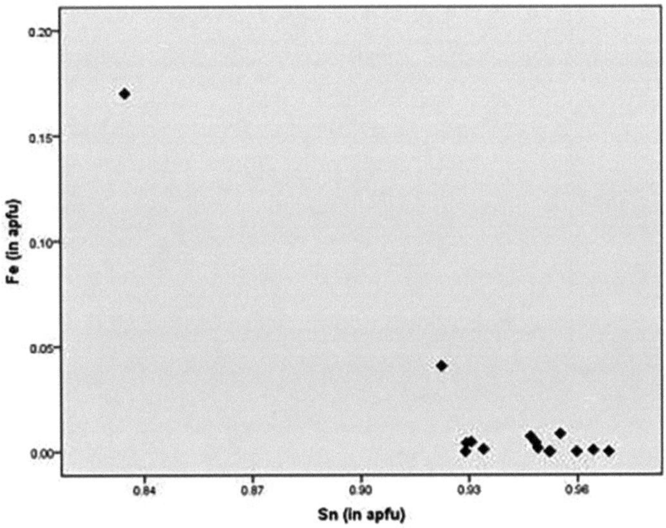

Figure 9.
Discrimination plot of Sn versus Fe (in apfu) in cassiterite of Musha-Ntunga.

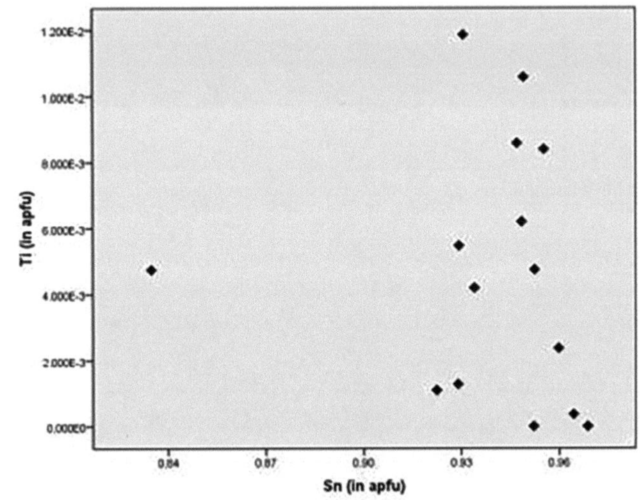

Figure 10.
Discrimination plot of Sn versus Ti (in apfu) in cassiterite of Musha-Ntunga.

A total of around 70 analyses were completed and the results are displayed in **Tables 9–11**. The cassiterite, wolframite and columbite-tantalite minerals in the investigated ore mineral samples are associated with silicate, oxide and sulfide mineral phases such as quartz, biotite, muscovite, garnet, tourmaline, Fe-minerals, rutile, ilmenite, zircon and sulfides (chalcopyrite, arsenopyrite and pyrite). This section will focus specifically on the mineral chemistry of the economically exploited ore minerals found in Bugarura-Kuluti: cassiterite, columbite-tantalite and wolframite.

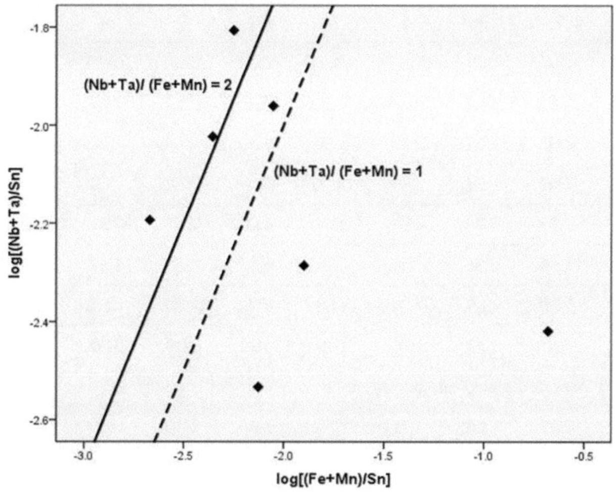

Figure 11.
Correlation diagram of selected cassiterite samples from Musha-Ntunga using Log [(Ta + Nb)/Sn] versus Log [(Fe + Mn)/Sn] after Möller et al. [18].

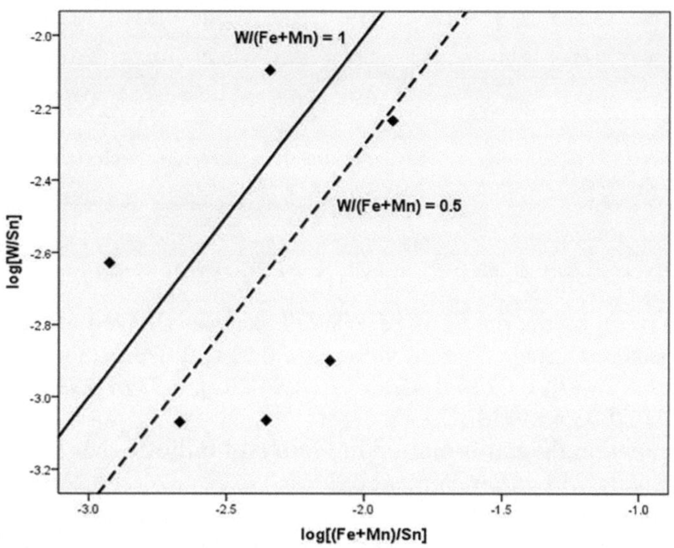

Figure 12.
Correlation diagram of selected cassiterite samples from Musha-Ntunga using log [W/Sn] versus log [(Fe + Mn)/Sn] after Möller et al. [18].

4.3.1 Mineral chemistry of cassiterite

Table 9 displays the cassiterite compositions indicating that majority of cassiterite grains are almost pure tin minerals, but that in some grains minor amounts of other mineral phases occur as inclusions.

The cassiterite mineralization is mostly composed of pure tin oxide – SnO_2 with average contents (per cassiterite sample) ranging between 90.50 and 99.08 ±1.3 wt. %.

Sample	1	2	3	4	5	6	7	8	9	10	Avg
Nb_2O_5	4.37	<DL	<DL	33.17	44.71	51.82	12.88	10.77	24.66	12.96	19.53
Ta_2O_5	73.34	77.88	84.7	45.75	33.24	27.69	75.4	74.7	60.3	73.73	62.68
Fe_2O_3	<DL	<DL	1.62	11.07	5.06	7.30	<DL	<DL	4.29	<DL	2.93
MnO	<DL	<DL	<DL	6.99	14.47	12.34	<DL	<DL	4.20	<DL	3.80
CaO	12.75	9.53	3.93	<DL	<DL	<DL	2.21	7.91	<DL	1.74	3.81
Na_2O	5.74	4.56	0.86	<DL	<DL	<DL	1.39	1.62	2.04	1.10	1.73
F	3.80	3.00	<DL	<DL	<DL	<DL	<DL	<DL	<DL	<DL	0.68
Tot.	100	100	100	100	100	100	100	100	100	100	100
Structural formula calculated based on six oxygen											
Nb	0.16	<DL	<DL	1.21	1.64	1.90	0.47	0.39	0.90	0.47	0.72
Ta	1.62	1.72	1.87	1.01	0.73	0.61	1.66	1.65	1.33	1.62	1.38
Fe	<DL	<DL	0.10	0.67	0.31	0.44	<DL	<DL	0.26	<DL	0.18
Mn	<DL	<DL	<DL	0.48	0.99	0.85	<DL	<DL	0.29	<DL	0.26
Ca	1.11	0.83	0.34	<DL	<DL	<DL	0.19	0.69	<DL	0.15	0.33
Na	0.90	0.72	0.13	<DL	<DL	<DL	0.22	0.25	0.32	0.17	0.27
Σcations	3.79	3.35	2.63	3.62	3.73	3.87	2.80	3.14	3.31	2.76	3.30
Nb/Ta	0.10	0.00	0.00	1.21	2.24	3.11	0.28	0.24	0.68	0.29	0.52
Fe/Mn	—	—	—	1.4	0.3	0.5	—	—	0.9	—	0.7

<DL: Below detection limit.
Total Fe- as Fe_2O_3 forms 2.93 wt. % on average, whereas the average Na content corresponds to 1.73 wt. % Na_2O. The sum of Nb + Ta is above 2 times all other cations and Ta > Nb (all in a.p.f.u).

Table 7.
Chemical composition of analyses of columbite-tantalite sample JC-25 from Musha-Ntunga.

In addition to tin oxide, the SEM-EDS/WDS analyses showed other major oxides for which respective average content values are 0.11 to 1.67±0.23 wt. % Nb_2O_5, 0.24 to 3.34±1.1 wt. % Ta_2O_5, 0.07 to 0.45±0.07 wt.% TiO_2, 0.37 to 5.9±0.2 wt.% Fe_2O_3 and 0.05 to 0.17±0.03 wt.% MnO.

The discrimination diagrams plotted in **Figure 14** indicate that tin oxide correlates poorly or negatively with other element oxides.

With the aim to understand the chemical variation of analyzed cassiterite samples, the results of analyses also were plotted in two variation diagrams, which are log [(Ta + Nb)/Sn] versus log [(Fe + Mn)/Sn] and log [W/Sn] versus log [(Fe + Mn)/Sn] after Möller et al. [18] (**Figures 15** and **16**).

On the one hand, the value distribution shows two types of samples (**Figure 14**): those which are plotting along (Nb + Ta)/(Fe + Mn) = 27 and others which are along (Nb + Ta)/(Fe + Mn) = 280. All analyzed samples plot above the line (Ta + Nb)/(Fe + Mn) = 2 (proportion 1:2) and, therefore, indicate that the cassiterite contamination is not controlled by a (Fe, Mn) (Ta, Nb)$_2$ O$_6$ mineral phase.

The excess of (Nb + Ta) content may point toward the existence of a simple substitution mechanism in the solid solution of (Nb, Ta) O_2 in SnO_2.

On the other hand, **Figure 16** shows that three types of substitution mechanisms in the Bugarura-Kuluti cassiterite occur in the W/Sn versus (Fe + Mn)/Sn discrimination diagram [18, 22]:

Analysis	Structural formula	Other chemical components detected by SEM-EDS	Possible group
1	$Na_{0.9}Ca_{1.1}Nb_{0.2}Ta_{1.6}O_6$	3.8 wt. % F	Tantalite
2	$Na_{0.7}Ca_{0.8}U_{0.1}Ta_{1.7}O_6$	3.0 wt. % F	Tantalite
3	$Na_{0.1}Ca_{0.3}Ba_{0.1}Fe_{0.1}U_{0.1}Ta_{1.9}O_6$	—	Tantalite
4	$Si_{0.2}Mn_{0.5}Fe_{0.7}Nb_{1.2}Ta_1O_6$	—	Ferro-columbite
5	$Mn_1Fe_{0.3}Ti_{0.01}W_{0.04}Ta_{0.7}Nb_{1.6}O_6$	0.15 wt. % SnO_2 and 0.38 wt. % UO_2	Mangano-columbite
6	$Si_{0.1}Mn_{0.8}Fe_{0.4}Nb_{1.9}Ta_{0.6}O_6$	—	Mangano-columbite
7	$Ba_{0.3}Na_{0.2}Ca_{0.2}Nb_{0.5}Ta_{1.7}O_6$	—	Tantalite
8	$Ba_{0.2}Na_{0.3}Ca_{0.7}Nb_{0.4}Ta_{1.6}O_6$	—	Tantalite
9	$Ba_{0.1}Na_{0.3}K_{0.1}Mn_{0.3}Fe_{0.3}Nb_{0.9}Ta_{1.3}O_6$	—	Columbite-tantalite
10	$Ba_{0.3}Na_{0.2}Ca_{0.2}Nb_{0.5}Ta_{1.6}O_6$	—	Tantalite

Table 8.
Structural chemical formula and nomenclature of selected "columbite-tantalite" analyses from Musha-Ntunga.

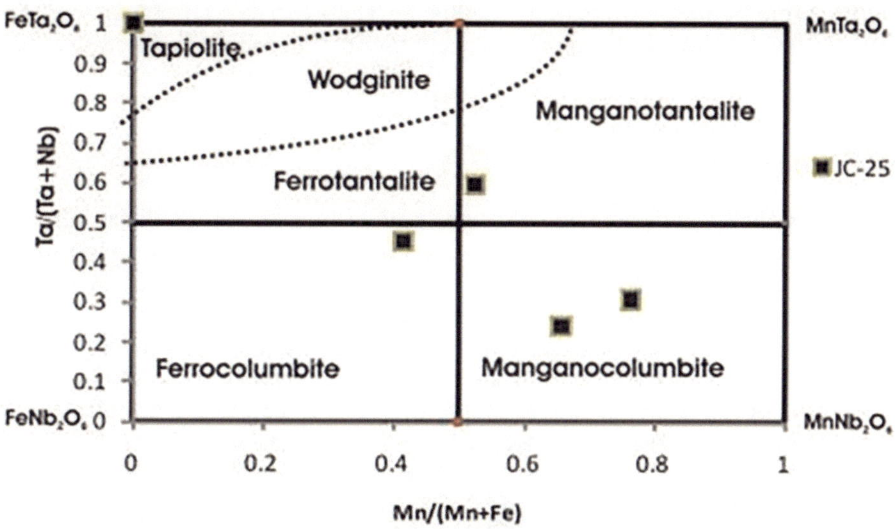

Figure 13.
Chemical composition of selected columbite-tantalite analyses from Musha-Ntunga represented in the FeTa-FeNb-MnNb-MnTa quadrilateral after Beurlen et al. [20].

- Type: $Sn^{4+} \leftrightarrow W^{4+}$ concerns mainly the samples of the first group (indicated in **Figure 16** as cass1) displaying excess of W over Fe + Mn. This should therefore compose with a solid solution of WO_2 in SnO_2 based on a simple substitution mechanism Sn-W.

- Type: $Sn^{4+} \leftrightarrow Fe^{3+} + H^+$ or $Sn^{4+} + O^{2-} \leftrightarrow Fe^{3+} + OH^-$ is associated with the group of cassiterite samples indicated as cass2 with a clear excess of Fe + Mn over the W

Elem/Samples	JC-33	JC-34	JC-28-6-1641	JC-28-6-2640	JC-28-6-4645	LLD
Nb_2O_5	<DL	1.69±0.22	<DL	0.23±0.09	n.d.	0.16
Ta_2O_5	<DL	3.34±1.10	<DL	<DL	n.d.	0.76
SnO_2	92.89±1.18	91.52±1.73	95.90±1.25	90.50±1.30	99.08±1.29	2.70
TiO_2	0.16±0.05	<DL	0.34±0.14	0.45±0.07	0.22±0.11	0.16
SiO_2	2.08±0.04	0.91±0.05	n.d.	1.85±0.05	n.d.	0.06
Fe_2O_3	<DL	0.37±0.06	2.38±0.14	5.91±0.21	n.d.	0.18
Al_2O_3	0.11±0.01	1.02±0.03	0.47±0.03	0.33±0.02	n.d.	0.04

Abbreviations: <DL-Below detection limit; LLD: lower limit of detection; n.d.: not determined.

Table 9.
Average chemical compositions of cassiterite group minerals from Bugarura-Kuluti.

content and according to Möller et al. [18], this may indicate the presence of a component such as Fe_2WO_6 in a solid solution or Fe-rich phases within cassiterite mineralization.

- Type: 2 $Sn^{4+} \leftrightarrow W^{6+} + (Fe, Mn)^{2+}$ related to the samples of a group named cass3 in the same figure plotting along the line W/(Fe + Mn) = 1 (proportion 1:1 on atomic level) which indicates that the contamination within these cassiterite samples seems to be likely controlled by solid inclusions of type (Fe, Mn) WO_4.

4.3.2 Mineral chemistry of columbite-tantalite

A study of chemical mineralogy of the columbite-tantalite mineralization from Bujumu was carried out on the only columbite-tantalite sample JC-34. Investigated areas on the sample were photographed by back-scattered electron image (BSE), structural formulae calculated and the results per analyses were co-plotted with the results of investigated samples from Rwanda after Melcher et al. [21] in the "columbite quadrilateral" after Beurlen et al. [20]. All the analytical data are reflected in **Table 10**.

The results of the analyses (**Table 10**) show that the "columbite-tantalite" mineral is mostly composed of tantalum, averaging 41.10+/−2.5 wt. % Ta_2O_5, niobium, with 36.50 +/−0.9 wt. % Nb_2O_5 in average, Fe_2O_3 with 9.93+/−0.7 wt. % and manganese with averages of 7.90+/−0.66 wt. % of MnO.

The W- average content is 1.51+/−0.7 wt. % WO_3, whereas the average Ti content corresponds to 0.39 +/−0.06 wt. % TiO_2. The bulk structural formulae and the main group in which those Nb-, Ta- bearing minerals of the analyses are compiled in **Table 11**.

The averages of Nb/Ta ratios of the columbite-tantalite sample from the study area are >1 which indicates enrichment in Nb. In addition, the Nb/Ta ratio > 1 allows us classifying this sample in the columbite group [23].

Lastly, and based on Beurlen et al. [20], the results of each analysis were reported in the quadrilateral diagram representing the end-member species in **Figure 17**. The columbite-tantalite sample from Bugarura-Kuluti shows Ta/(Ta + Nb) ratio varying between 0.382 and 0.439 and a constant value of Mn/(Mn + Fe) ratio which is 0.47. In the Beurlen et al. [20] diagram, they are plotting in the ferro-columbite group rich end-members. In comparison with Melcher et al. [21], the Bugarura-Kuluti columbite-

DOI: http://dx.doi.org/10.5772/intechopen.1008342

Bugarura-Kuluti*					Cyubi**			Ruhanga**			Gasasa**	
	1	2	3	LLD*	6	8	10	1	100	102	20	21
WO_3	2.55	1.08	0.89	2.02	n.d.	n.d.	0.114	0.128	n.d.	n.d.	0.236	0.161
Nb_2O_5	34.4	39.72	35.38	1.84	19.32	5.57	5.54	59.09	14.01	4.37	47.96	26.07
Ta_2O_5	44.68	40.88	37.74	5.83	62.93	77.19	63.96	19.45	68.69	80.02	32.46	56.36
TiO_2	1.18	n.d.	n.d.	0.12	0.18	0.229	2.54	0.654	0.255	n.d.	0.253	0.205
SnO_2	0.79	n.d.	1.28	0.52	n.d.	0.586	13.1	0.17	0.218	n.d.	n.d.	n.d.
Fe_2O_3	9.09	10.16	10.54	1.43	11.2	14.32	7.09	3.65	8.53	1.97	12.75	13.56
MnO	7.31	8.16	8.24	1.32	5.54	1.48	6.52	15.69	7.05	12.34	6.82	4.09
Total	100	100	100		99.17	99.37	98.86	98.83	98.753	98.7	100.48	100.4
Structural formula calculated on the basis of six oxygens												
W	0.044	0.019	0.015	n.d.	0.000	0.000	0.002	0.002	0.000	0.000	0.004	0.003
Nb	1.036	1.196	1.065	n.d.	0.582	0.168	0.167	1.780	0.422	0.132	1.444	0.785
Ta	0.810	0.741	0.684	n.d.	1.141	1.399	1.159	0.353	1.245	1.450	0.588	1.021
Ti	0.059	0.000	0.000	n.d.	0.009	0.011	0.127	0.033	0.013	0.000	0.013	0.010
Sn	0.021	0.000	0.034	n.d.	0.000	0.016	0.347	0.005	0.006	0.000	0.000	0.000
Fe	0.455	0.509	0.528	n.d.	0.561	0.717	0.355	0.183	0.427	0.099	0.638	0.679
Mn	0.412	0.460	0.465	n.d.	0.313	0.083	0.368	0.885	0.398	0.696	0.385	0.231
Σcat.	2.837	2.925	2.834	n.d.	2.005	2.394	2.525	3.23	2.510	2.377	3.072	2.729
Nb/Ta	1.279	1.615	1.558	n.d.	0.510	0.120	0.144	5.048	0.339	0.091	2.455	0.769
Nb + Ta	1.846	1.937	1.749	n.d.	1.722	1.567	1.326	2.132	1.667	1.582	2.033	1.807
Ta/(Ta + Nb)	0.439	0.38	0.391	n.d.	0.662	0.893	0.874	0.165	0.747	0.917	0.289	0.565
Mn/(Fe + Mn)	0.475	0.475	0.468	n.d.	0.358	0.104	0.509	0.829	0.482	0.876	0.376	0.254

Present study, LLD-lower limit of detection. *Melcher et al. [21] with LLD (in wt. %) of 0.03 (WO_3), 0.15 (Nb_2O_5), 0.12 (Ta_2O_5), 0.04 (TiO_2), 0.03 (SnO_2), 0.03 (Fe_2O_3) and 0.04 (MnO); n.d.-not determined.*

Table 10.
Chemical composition of columbite-tantalite analyses of sample JC-34 from Bugarura-Kuluti (this study) compared to those of the Cyubi, Ruhanga and Gasasa mines [21].

Analysis	Structural formula	Other chemical elements detected by SEM-EDS	Possible group
1	$Mn_{0.412}Fe_{0.455}Nb_{1.036}Ta_{0.81}O_6$	W, Ti and Sn	Ferro-columbite
2	$Mn_{0.46}Fe_{0.509}Nb_{1.196}Ta_{0.741}O_6$	W, Ti and Sn	Ferro-columbite
3	$Mn_{0.465}Fe_{0.528}Nb_{1.065}Ta_{0.684}O_6$	W, Sn, In? and Tl?	Ferro-columbite

Table 11.
Structural chemical formula and nomenclature of selected "columbite-tantalite" analyses from Bugarura-Kuluti.

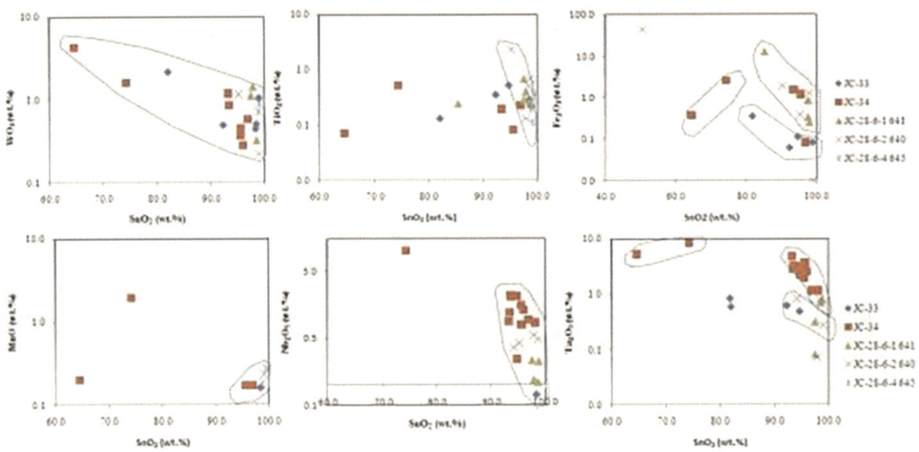

Figure 14.
Bivariate diagrams of SnO2 against WO3, TiO2, Fe2O3, Nb2O5, Ta2O5 and MnO (in wt.%) for the cassiterite samples from Bugarura-Kuluti.

tantalite is enriched in Fe and Nb compared to the samples from Cyubi and Gasasa (enriched in Ta and Fe) or from Ruhanga (enriched in Mn).

In spite of the more or less sparse distribution of the complete data set in this columbite quadrilateral, the general trends begin with Nb-rich compositions for less fractionated pegmatites and evolve to the Ta-rich series. According to Linnen and Keppler [24] and Beurlen et al. [20], this increase in Ta/(Ta + Nb) ratio with the degree of felsic magmatic melts fractionation should be explained as a consequence of the lower solubility of Nb-rich columbite group members in peraluminous granite/pegmatite melts in comparison with the Ta-rich end-members.

4.3.3 Mineral chemistry of wolframite from Bugarura-Kuluti

Ferberite ore samples, Lot JC-36 from Rwankuba (Bugarura-Kuluti-Bibare concession) have been obtained from the collection of RMCA and analyzed using SEM-EDS/WDS facilities at the Dept. Geol., UFS, BFN, RSA (**Table 12**).

Besides the tungsten ore, the paragenetic sequence also reveals the existence of quartz and Fe- minerals which may possibly correspond to hematite and/or goethite.

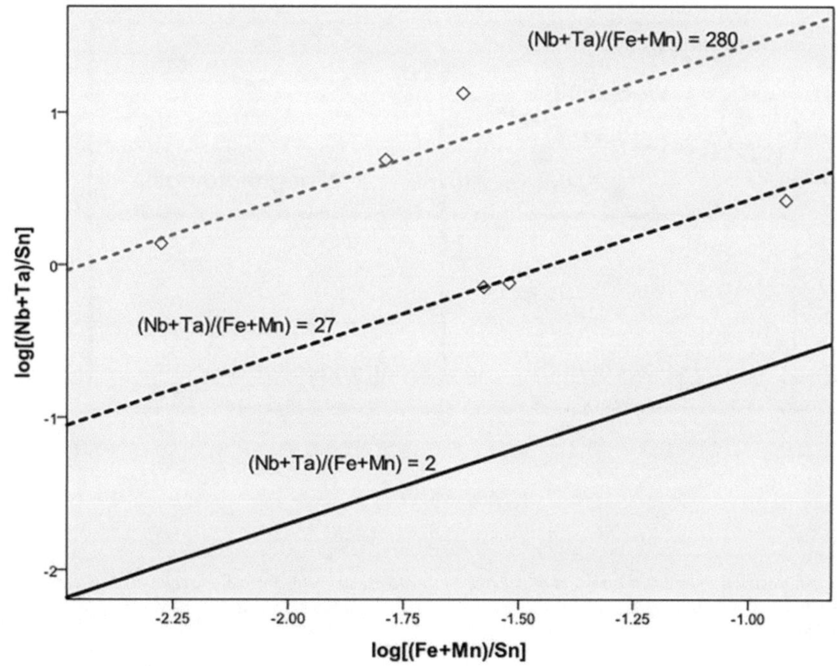

Figure 15.
Correlation diagrams of analyses for cassiterite samples from Bugarura using Log [(Ta + Nb)/Sn] versus Log [(Fe + Mn)/Sn] after Möller et al. [18]. Ta, Nb, Fe, Mn and Sn contents are in apfu.

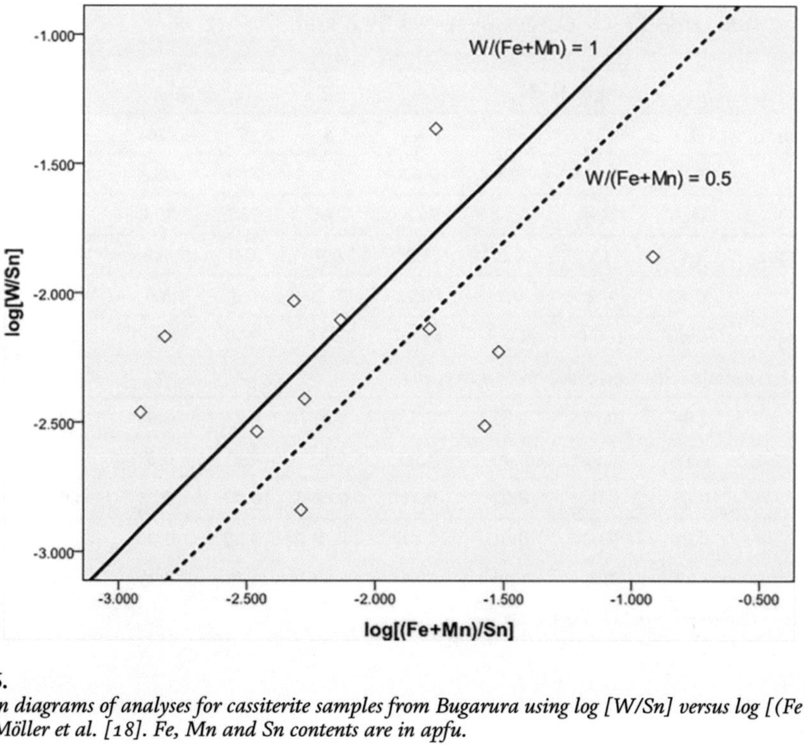

Figure 16.
Correlation diagrams of analyses for cassiterite samples from Bugarura using log [W/Sn] versus log [(Fe + Mn)/Sn] after Möller et al. [18]. Fe, Mn and Sn contents are in apfu.

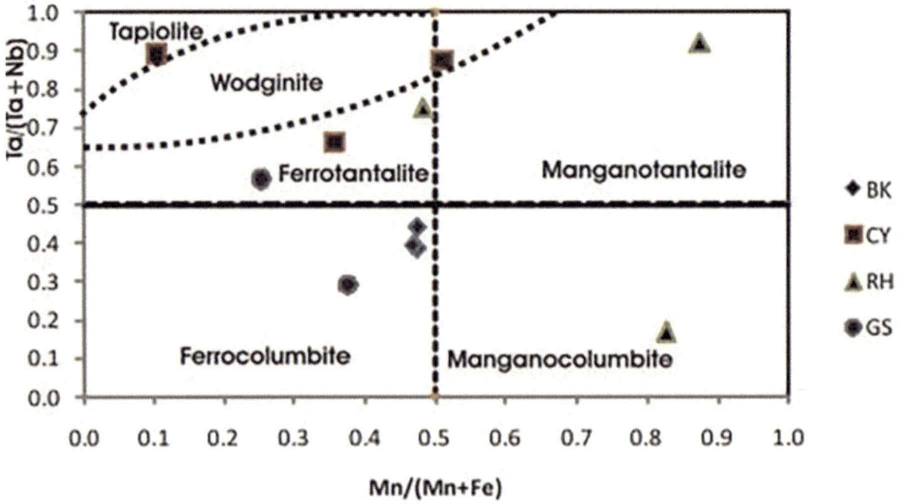

Figure 17.
Chemical composition of selected analyses of columbite-tantalite samples from Bugarura-Kuluti (BK) (this study) and from Cyubi (CY), Ruhanga (RH) and Gasasa (GS) [21] plotted in FeTa-FeNb-MnNb-MnTa quadrilateral after Beurlen et al. [20].

The sample seems likely to contain minor amounts of phosphate minerals. Cassiterite and columbite-tantalite are in traces.

The eight analyses of the wolframite samples show three dominant oxides:

1. *Tungsten oxide-WO_3:* WO_3 contents range from 67 to 75.3 wt. %.

2. *Total iron oxide-Fe_2O_3* ranges between 19.8 and 32.3 wt. %.

	Sample JC-36								
Analysis:	1	2	3	4	5	9	10	11	LLD
WO_3	71.3	74.2	72.5	70.2	70.8	71.9	75.3	67.0	3.6
Fe_2O_3	24.7	24.4	25.2	26.4	25.6	23.7	19.8	32.3	0.7
MnO	1.4	1.1	0.5	3.0	0.9	0.9	3.9	<DL	0.3
MgO	0.4	0.2	0.2	0.2	0.2	n.d.	n.d.	0.5	0.1
Total	100.0	100.0	100.0	100.0	100.0	99.9	100.0	100.0	
Structural formula calculated based on four oxygens									
W	0.84	0.87	0.85	0.82	0.83	0.84	0.88	0.79	—
Fe	0.84	0.83	0.86	0.90	0.87	0.81	0.67	1.10	—
Mn	0.05	0.04	0.02	0.11	0.04	0.03	0.15	0.00	—
Mg	0.02	0.02	0.01	0.01	0.01	0.00	0.00	0.03	—
∑Cat.	1.82	1.76	1.80	1.85	1.84	1.78	1.73	1.92	—

Abbreviations: LLD-lower limit of detection.

Table 12.
Summary of the chemical composition of investigated tungsten-bearing ore samples from Bibare mining concession (Rwankuba mine).

3. *Manganese oxide-MnO:* The measured contents are from below 1 up to 3.9 wt. %.

The calculated structural chemical formula is almost the same for all the analyses and can roughly be written as follows: $FeWO_4$ which indicates that the investigated wolframite samples are iron-rich end-members (ferberites).

4.4 Fluid evolution

The results from the fluid inclusion in quartz studies show that:

- Most of the investigated inclusions are two phases (liquid–vapor). They are liquid-dominated but few of them are vapor-dominated, indicating that the boiling phase traces were overprinted by later tectonic events or the phase itself was less important. In addition, monophase and a limited number of multiphase inclusions were also identified.

- Total melting temperatures of ice range between −0.3 °C and − 13.5 °C, representing the salinities between 0.5 wt. % NaCl equivalent and 17.3 wt. NaCl equivalents (**Figure 18**).

- The homogenization temperatures recorded were between 103°C and 360°C.

See **Figure 18**.

Insofar as the depth of the Rugarama granite is estimated to be 7.5–8 km, indicating lithostatic pressure values of more or less 2kbars [25] and using the equation of Bodnar and Vityk [26], the trapping temperature estimates in accordance with these previously mentioned conditions were between 150°C and 560°C for the quartz samples from Bugarura-Kuluti, 220°C and 320°C for the Rwinkwavu quartz samples and

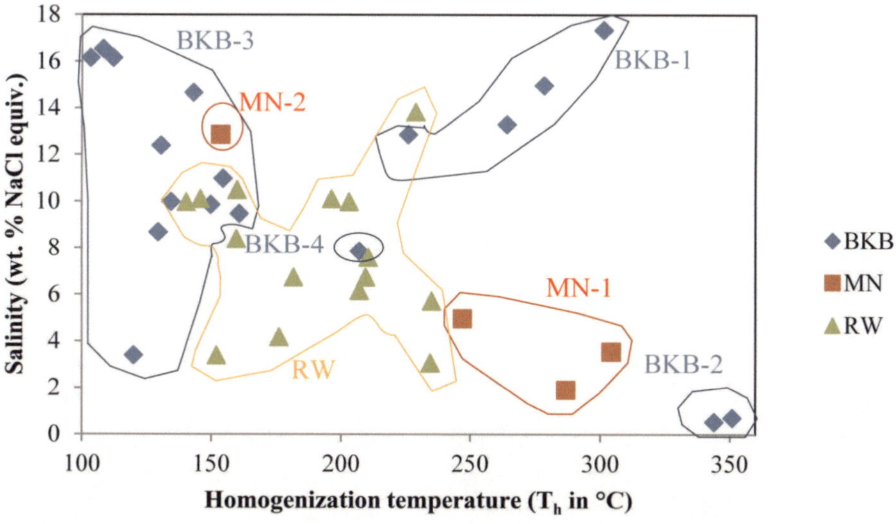

Figure 18.
Bivariate diagram of homogenization temperatures (Th in °C) versus salinities (in wt. % NaCl equivalent) for the two-phase fluid inclusions in the studied samples. Abbreviations: BKB-Bugarura-Kuluti-Bibare; MN-Musha-Ntunga; RW-Rwinkwavu.

between 200 °C and 420°C for the samples from Musha-Ntunga. The high values of pressure-corrected temperatures obtained by fluid inclusion micro-thermometry on quartz samples from Musha-Ntunga (this study) overlap with the formation temperature of cassiterite (≈ 400°C) based on the isotopic equilibrium between cassiterite and quartz from Musha [2].

In summary, the results of fluid inclusion investigations show a wide range of salinities and homogenization temperatures. There should be only two possible explanations for this: by the presence of different hydrothermal populations, which overlap or this might be due to a mixing of magmatic solutions with non-magmatic fluids. The latter can be composed of meteoric, connate and/or metamorphic waters. The stable isotope study on mineralized quartz by Dewaele et al. [2] confirmed that the mineralizing fluids of Musha are from the interaction between primary magmatic fluids and metamorphic waters. To extend this hypothesis on the origin and evolution of the mineralizing fluids in the Bugarura-Kuluti-Bibare and Rwinkwavu prospects, further stable isotope investigations are recommended. The fluid-inclusions homogenizing at low temperatures may represent the younger generation of secondary inclusions. The fluid-inclusions which homogenize at moderate to high temperatures might be older and could be similar to the primary inclusions which have all disappeared due to a long-lived magmatic and tectonic history in the Kibaran metallogenetic province.

5. Discussions

5.1 Mineral precipitation conditions as reflected by cassiterite, wolframite and columbite-tantalite chemistry

On the one hand, it is confirmed that the cassiterite of the study area is associated with the late to post-magmatic water and volatile-rich residual melts (e.g., this study; [2, 8]) and in hydrothermal solutions, tin is mostly transported as reduced Sn^{2+} chloro-complexes (e.g. $SnCl_2^0$) rather than as Sn^{4+} hydroxy-chloro-complexes and the precipitation of cassiterite involves oxidation to Sn^{4+} [27]. This latter author also showed that three possible mechanisms were responsible for the cassiterite deposition:

- Vapor separation, which removes HCl and hence causes the precipitation of cassiterite as illustrated by the following chemical equation: $SnCl_2 + H_2O + 0.5\ O_2 = SnO_2 + 2H^+ + 2Cl^-$,

- Mixing of hot saline magmatic melts with non-magmatic fluids during emplacement of granites and subsequent metasomatic process and

- Acid neutralization by feldspar hydrolysis that takes place during the greisenization process after [$Sn^{2+} + 3(Na, K)\ AlSi_3O_8 + 2H_2O = SnO_2 + KAl_3Si_3O_{10}(OH)_2 + 6\ SiO_2 + 2Na + H_2$] equation.

On the other hand, log [(Fe + Mn)/Sn] versus log [Nb + Ta)/Sn] (upper part of **Figure 19**) and log [(Fe + Mn)/Sn] versus log [W/Sn] (lower part of the same **Figure 19**) discrimination plots [18] for the cassiterite samples from the study area together with the cassiterite samples from Egyptian Pan-African Orogeny [22] may give indication of petrogenetic conditions.

Some of the cassiterite samples from Rwinkwavu, Musha-Ntunga (this study) are showing roughly a positive trend plotting along the line (Nb + Ta)/(Fe + Mn) = 2 in the first diagram, indicating a contamination by mineral inclusions of the type (Fe, Mn)(Ta, Nb)$_2$O$_6$, but the rest of the samples from the study area show enrichment trends in (Fe + Mn) which may indicate the existence of Fe-rich phases by simple substitution of Sn by Fe.

Contrarily, all the samples from Bugarura-Kuluti (part of this study) plot above (Nb + Ta)/(Fe + Mn) = 2 and thus are (Nb + Ta)-dominated. This chemical trend may indicate that there is a substitution of the type $Sn^{4+} \leftrightarrow (Ta, Nb)^{4+}$ in the cassiterite crystals.

Some cassiterite samples from the study area (Bugarura-Kuluti, BK; Musha-Ntunga, MN) are plotting along the line W/(Fe + Mn) = 1 and according to Möller et al. [18], such ratios toward the presence of wolframite in cassiterite as solid inclusions. The rest of the samples, including the whole set from Rwinkwavu, show excess in Fe and are therefore indicating the presence of Fe-rich mineral phases (**Figure 19**).

With regard to the tungsten mineralization, tungsten is a hard metal existing in nature in the form of W^{6+} or W^{5+} and, therefore, complexing with hard bases. In hydrothermal solutions, W species are transported in the form of tungstates [28]. The stability of tungstates is Eh and pH-dependent [29] and therefore, the W- ore precipitates in reducing conditions rather than the oxidizing ones.

The chemical analysis of the wolframite samples indicates that it is in a pure ferberite type with the chemical formula $FeWO_4$ and other chemical constituents are in small amounts or in traces.

The mineral chemistry of columbite-tantalite mineralization was investigated and the results plotted (**Figure 20**) in columbite quadrilateral [20] binary plot of Mn/(Fe + Mn) versus Ta/(Ta + Nb) together with the columbite-tantalite samples from other deposits of Rwanda such as Ruhanga, Cyubi and Gasasa [21]. The samples show enrichment trends in Ta or Mn: Musha-Ntunga (MN) and Bugarura-Kuluti (BK)-trend 2 (this study) and for deposits-trend 3 [21]. Based on Pohl [23], these enrichment trends in Ta and Mn are an indication that the columbite-tantalite samples from Rwanda have crystallized from more evolved felsic magmas.

According to Linnen [30, 31], the solubility of Fe-rich members of the columbite group minerals in the melts is larger than that of Mn-rich end-members and the enrichment in Mn over Fe might be likely controlled by other Fe-bearing minerals crystallized during the pegmatite evolution such as tourmaline or biotite [32]. The more Fe-rich minerals are in paragenesis, the poorer in Fe the columbite is. In normal conditions, Fe-rich columbites represent the late stage of magmatic evolution and the increasing of Mn in this case might be dictated by a competition for Fe between the present minerals.

This study revealed the frequent occurrence of Fe-rich tourmalines, schorl, in concentric tourmalinite rims along the pegmatite/vein intrusions. This indicates that the B-enriched parent fluids were of magmatic origin [33]. The key role of B in the lowering of the granite solidus temperatures and increasing the range of temperatures over which magmatic crystallization may occur was documented in Dingwell [34]. London [35] and Pollard et al. [36] noted that B and other fluxing elements such as F, Li and P were responsible for the increase of H_2O-solubility in the magmas, causing the decrease of viscosity and the solidus temperatures of the magmatic melts which is

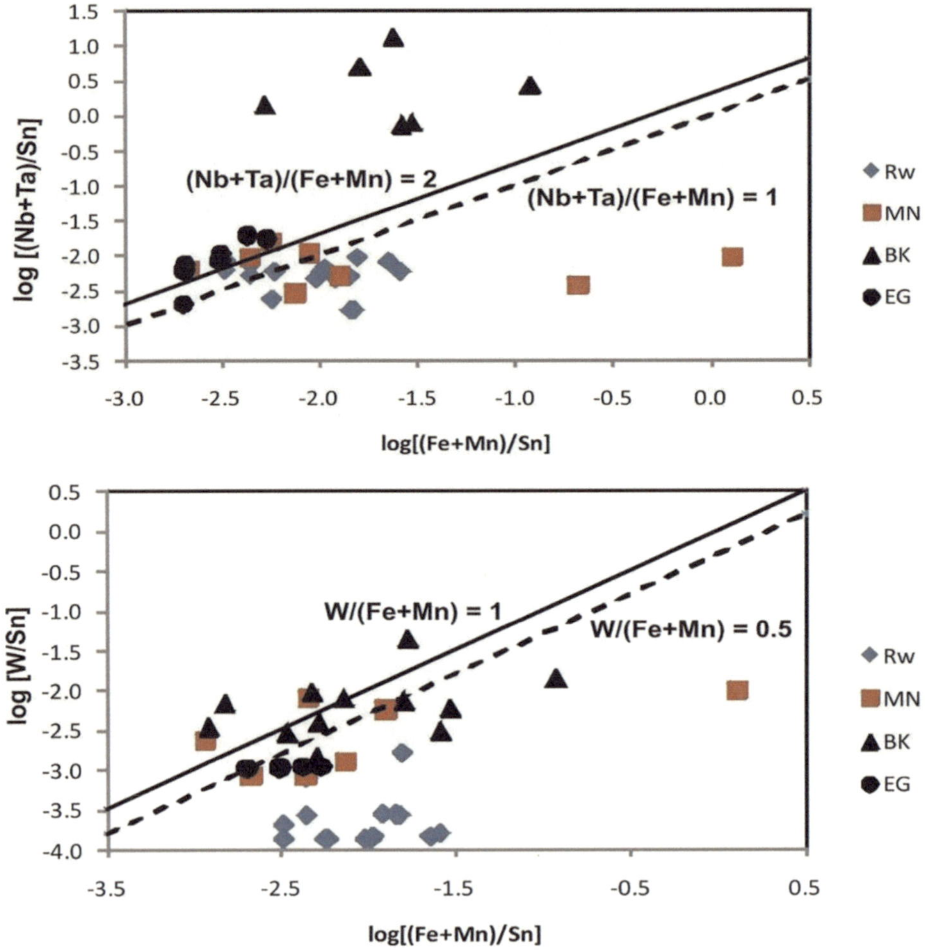

Figure 19.
Discrimination plots of selected cassiterite samples from the study area using log [(Nb + Ta)/Sn] versus log [(Fe + Mn)/Sn] (upper) and log [W/Sn] versus log [(Fe + Mn)/Sn] (lower) (after [18]) in comparison with cassiterite samples from granites of the Egyptian Pan-African Orogeny [22]. Abbreviations: Rw-Rwinkwavu, MN-Musha-Ntunga, BK-Bugarura-Kuluti, EG-Egypt. The concentrations are in apfu.

in favor of crystal-melt separation. This, therefore, promotes rare metal enrichments in residual melts. In the Bugarura-Kuluti and Musha-Ntunga prospects, the assumed simultaneous increase of the water-solubility in the residual melts and the enrichment of hydrothermal solutions in Sn, W and probably other incompatible elements, might have generated an increase of a vapor-dominated phase with relatively high salinities and possibly richer in Nb and Ta than in Sn and W. The results of fluid inclusion investigations on quartz samples from Bujumu showed that a few inclusions, possibly equivalent to the remnants of the early generation of inclusions, are vapor-dominated, more saline than others, with relatively high T_h (250°C - 300°C) and thus, tend to have a "pneumatolytic" character.

These conditions are likely in favor of pegmatite-type columbite-tantalite-dominated mineral precipitation in some known mine sites of Bugarura-Kuluti and Musha-Ntunga. This hypothesis is consistent with Linnen [37] who showed that the Nb and Ta were more associated with silicate melts rather than hydrothermal aqueous fluids.

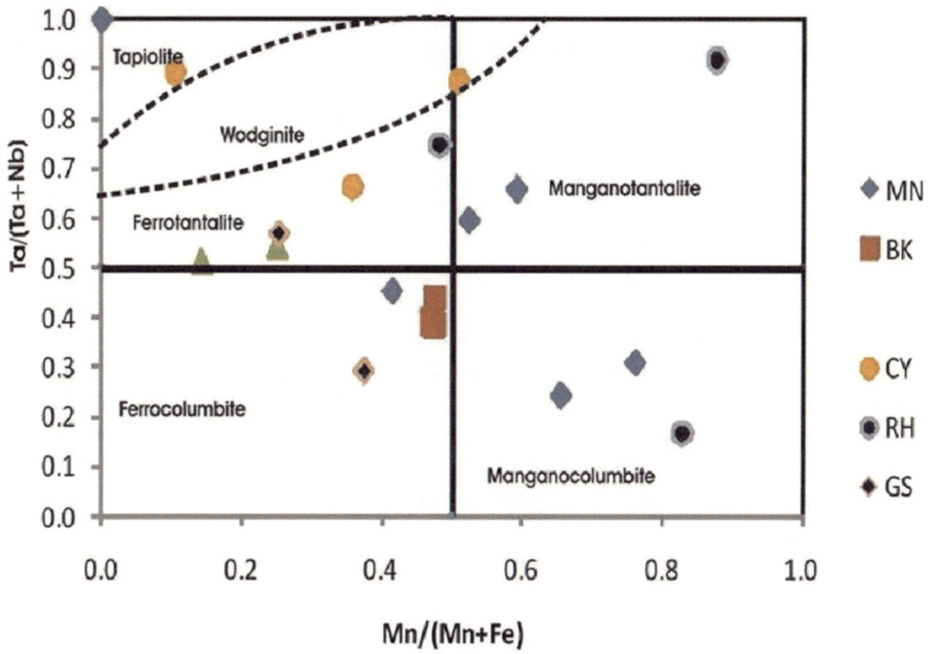

Figure 20.
Columbite quadrilateral binary plot of Mn/(Fe + Mn) versus Ta/(Ta + Nb) in a.p.f.u [20] for the columbite-tantalite samples from the study area and from the selected mines in Rwanda [21]. Abbreviations: MN-Musha-Ntunga, BK-Bugarura-Kuluti, CY-Cyubi, RH-Ruhanga and GS-Gasasa, 1, 2 and 3-evolution trends.

Fetherston [38] showed in experiments that the Ta starts to react mostly with other chemical components at temperatures above 300 to 400°C and remains immune to chemical attacks on the temperatures below 150°C. Cerny et al. [39] supported by Burnham [40], considered the fluid saturation as a consequence of hydrous melts precipitating anhydrous minerals such as quartz and feldspars. The decrease in water solubility may be considered as a consequence of decreasing pressure as the granite magmas rise through the crust during the late stages of fractional crystallization. The Sn might have been mobilized and transported by the residual ascending Cl-bearing and water-rich aqueous fluids exsolved at shallow crustal levels [27, 41]. The fluid oversaturation in Sn and precipitation of cassiterite in hydrothermal veins might have been derived from combined processes of decrease in temperature and/or the chloride activity. The influx of cold meteoric waters in the ascending hydrothermal fluids influenced the increase of oxygen fugacity and/or pH. This is evidenced by a wide spectrum of T_h (103–360°C) of fluid inclusions. The mixing of fluids from different sources is also supported by the contrasts of salinities (0.5–17.3 wt. % NaCl equiv.) between various investigated inclusions from the study area. Moreover, the active role of the interaction between the resulting fluids and the host rocks causing destabilization of complexes and the cassiterite precipitation may not be excluded [42]. In the study area the cassiterite precipitated possibly either according to the $SnCl_2 + H_2O + 0.5O_2 = SnO_2 + 2H^+ + 2Cl^-$ equation which is Eh and pH-dependent and is followed by acidification or the acid neutralization by hydrolysis of feldspars of the wall - rocks as follow: $Sn^{2+} + 3\ (Na, K)\ AlSi_3O_8 + 2\ H_2O = SnO_2 + KAl_3Si_3O_{10}\ (OH)_2 + 6\ SiO_2 + 2Na + H_2$ [27]. For this latter reaction, the final products include not only cassiterite but also muscovite and quartz.

The precipitation of ferberite was possibly caused by oversaturation in Fe of the resulting fluids from the mixture between the W-rich and Fe-bearing hydrothermal fluids of magmatic origin. This is confirmed by the fluid inclusion investigations as well as meteoric (connate/metamorphic) waters with high Fe content (leached from the host metasediments) [43]. The Fe of magmatic origin together with the Fe dissolved in the surface cold waters might have reacted together with W to form the ferberite.

Reducing conditions caused by carbonaceous matter in the metasediments played a key role in the ferberite deposition.

The anomalous metal distribution patterns of meta-sedimentary rocks in the study area might have been originated by rare elements-rich ascending hydrothermal fluids. Similar cases were noted by Beer and Ball [44] in the pelitic rocks extended around the contact zones of granites and in mineralized areas of SW-England introducing metasomatic fluids in the host rocks and causing Sn and W enrichments. The results from fluid inclusion study on cassiterite-bearing pegmatite/vein quartz samples from Rwinkwavu showed that the cassiterite and ferberite mineralization is more abundant in aqueous, less saline and low temperature hydrothermal fluids.

6. Conclusions

From the present research project, it can be concluded that the mesoproterozoic meta-sedimentary rocks of South-Eastern Rwanda were intruded by two granite generations: the oldest G1–3 of roughly 1380±10 Ma age and the younger G4- granites of 986±10 Ma age [8]. The latter is associated with the Sn-W-Nb-Ta-(Au) mineralization which occurs in the three investigated mineral deposits (Rwinkwavu, Musha-Ntunga and Bugarura-Kuluti) [2, 8, 15, 45].

The primary mineralization is pegmatite/hydrothermal quartz veins type and tectonically controlled. In addition, the secondary deposits exist in the form of alluvial/eluvial placers. The salinities of the mineralizing fluids are between 0.5 and 17.3 wt. % NaCl equivalent. The trapping temperatures are extending from 150°C to above 500° C. This contrast of salinities and the wide range of formation temperatures between various investigated fluid inclusions might be due to a mixing of magmatic fluids with metamorphic to meteoric waters.

The mineral associations in the Eastern Rwandan ore deposits are columbite-tantalite-muscovite-quartz-tourmaline, cassiterite-sericite-muscovite-tourmaline and ferberite-huebnerite-quartz-muscovite-garnet-ilmenite-hematite-goethite.

In the cassiterite samples from Bugarura-Kuluti, the dominant substitution is Sn^{4+} replaced by $(Ta, Nb)^{4+}$. In the cassiterites from Musha-Ntunga and Rwinkwavu prospects, three Sn^{4+} cations are replaced by two $(Ta, Nb)^{5+}$ + $(Fe, Mn)^{2+}$ cations, and/or Sn^{4+} + O^{2-} is replaced by Fe^{3+} + OH^{-}. The cassiterite samples also show intergrowths with W-rich mineral phases.

The columbite-tantalite samples from Rwanda show increased incompatible element contents ([21, 23] and this study). This is an indication that they have crystallized from more evolved felsic magmas.

In terms of tungsten mineralization, the dominant wolframite mineral type is ferberite.

Different generations of fluid inclusions were found to exist, as well as ore mineral zoning and various types of hydrothermal alteration. This indicates that the metallogenetic evolution in the eastern Rwandan mineral deposits took place in multiphases involving multi-stage circulation of hydrothermal fluids.

7. Highlights

- The lithology of South-Eastern Rwanda is composed of meta-sedimentary rocks of Mesoproterozoic age, which were intruded by two granite generations: the oldest G1–3 of roughly 1380±10 Ma age and the younger G4- granites of 986±10 Ma age [8]. The latter is associated with the Sn-W-Nb-Ta-mineralization, which occurs in the three investigated mineral deposits (Rwinkwavu, Musha-Ntunga and Bugarura-Kuluti) [2, 8, 15].

- The primary mineralization is pegmatite/hydrothermal quartz vein type and is tectonically controlled. In addition, the secondary deposits exist in the form of alluvial/eluvial placers. The salinities of the mineralizing fluids are between 0.5 and 17.3 wt.% NaCl equivalent. The trapping temperatures are from 150°C to above 500°C. The contrast of salinities and the wide range of formation temperatures between various investigated inclusions might be due to a mixing of magmatic fluids with metamorphic to meteoric waters.

- The mineral associations in the eastern Rwandan ore deposits are columbite-tantalite-muscovite-quartz-tourmaline, cassiterite-sericite-muscovite-tourmaline and ferberite-huebnerite-quartz-muscovite-garnet-ilmenite-hematite-goethite.

- In the cassiterite samples from Bugarura-Kuluti, the dominant substitution is Sn^{4+}, replaced by $(Ta, Nb)^{4+}$. In the cassiterites from Musha-Ntunga and Rwinkwavu prospects, the three Sn^{4+}, is replaced by $2(Ta, Nb)^{5+}$ + $(Fe, Mn)^{2+}$ and/or Sn^{4+} + O^{2-} is replaced by Fe^{3+} + OH^{-}. The cassiterite samples also show intergrowths with W-rich mineral phases.

- The columbite-tantalite samples from Rwanda show increased incompatible element contents ([21, 23] and this study). This is an indication that they have crystallized from more evolved felsic magmas.

- The dominant wolframite mineral is ferberite.

- There are different generations of fluid inclusions, ore mineral zoning and various types of hydrothermal alterations. This indicates that the metallogenetic evolution in the eastern Rwandan mineral deposits took place in several phases involving multi-stage circulation of hydrothermal fluids.

Acknowledgements

Parts of this chapter were previously published in the thesis by the same author: (Jean Claude Ngaruye). Genetic characterization of (Petrographic and geochemical investigation of Sn - W - Nb - Ta pegmatites and mineralized quartz veins in South-Eastern Rwanda) [45]. (Unpublished MSc thesis referenced in the current chapter). (University of the Free State). Available from: https://www.semanticscholar.org/paper/Petrographic-and-geochemical-investigation-of-Sn-W-Ngaruye/ea33521a0370a7202745d0baec3e8354c07b25cf.

The authors would like also to thank anonymous reviewers for their constructive comments and suggestions.

Author details

Jean-Claude Ngaruye[1*] and Christoph Gauert[2]

1 Rwanda Mines, Petroleum and Gas Board, Kigali, Rwanda

2 Department of Geology, Geological Survey of Saxony, Germany

*Address all correspondence to: jclaude.ngaruye@gmail.com

References

[1] OGMR. Mineral Deposits Map of Rwanda. Vol. 12009. p. 250000

[2] Dewaele S, De Clerq F, Muchez P, Schneider J, Burgess R, Boyce A, et al. Geology of the cassiterite mineralization in Rutongo area, Rwanda (Central Africa): Current state of knowledge. Geologica Belgica. 2010;**13**(1):91-112

[3] Dewaele L, Fernandez-Alonso M, Boyce A, Muchez P. Cassiterite and columbite - tantalite mineralization in pegmatites of the northern part of the Kibaran orogen (Central Africa): The Rutongo area (Rwanda). In: Andrew CJ et al., editors. Proceedings of the 9th Biennal SGA Meeting. Mineral Exploration and Research: Digging Deeper; 20th–23rd August 2007, Dublin, Ireland. 2007a. pp. 1489-1492

[4] Dewaele S, Tack L, Fernandez-Alonso M, Boyce A, Muchez P. Cassiterite mineralization in vein-type deposits of the Kibaran orogen (Central Africa): Nyamyumba (Rutongo area, Rwanda). In: Andrew CJ et al., editors. Proceedings of the 9th Biennial SGA Meeting. Mineral Exploration and Research: Digging Deeper 20th–23rd August 2007, Dublin, Ireland. 2007b. pp. 1007-1010

[5] Dewaele S, Tack L, Fernandez-Alonso M, Boyce A, Muchez P, Schneider J, et al. Geology and mineralization of the Gatumba area, Rwanda: Present state of knowledge. Etudes Rwandaises. 2008;**16**: 6-24

[6] Pohl W. Metallogeny of north-eastern Kibara belt, Central Africa-recent perspectives. Ore Geology Reviews. 1994;**9**:105-130

[7] Pohl W, Günther MA. The origin of Kibaran (late mid-Proterozoic) tin, tungsten and gold quartz vein deposits in Central Africa: A fluid inclusions study. Mineralium Deposita. 1991;**26**:51-59

[8] Tack L, Wingate MTD, De Waele B, Meert J, Belousova E, Griffin B, et al. The 1375 ma "Kibaran event" in Central Africa: Prominent emplacement of bimodal magmatism under extensional regime. Precambrian Research. 2010; **180**:63-84

[9] Makitie H, Data G, Isabirye E, Mantari I, Huhma H, Klausen MB, et al. Petrology, geochronology and emplacement model of the giant 1.34 Ga Lake Victoria dyke swarm of the margin of a large igneous province in eastern Africa. Journal of African Earth Sciences. 2014;**97**:273-296

[10] Van Daele J, Scherer E, E. Neoproterozoic pre- and post-deformational metamorphism in the Western domain of the Karagwe-Ankole Belt reconstructed by Lu-Hf garnet geochronology in the Kibuye-Gatumba area, Rwanda. Precambrian Research. 2020;**344**

[11] Tack L, Fernandez-Alonso M, De Waele B, Tahon A, De Waele S, Baudet D, et al. The northeastern Kibaran Belt (NKB): A long-lived intraplaque history. In: Extended Abstract of Oral Communication Submitted to Session 1 ("Geodynamics of Africa") of the 21st Colloquium of African Geology (CAG21) at Maputo (Mozambique), 3–5 July 2006. 2006. pp. 149-151

[12] Tack L, Wingate M, Dewaele B, Meert J, Griffin B, Belousova EA, et al. The Proterozoic Kibaran Belt in Central Africa: Intra-cratonic 1375 ma emplacement of a LIP. In: Abstract for the 22nd Colloquium African Geology (CAG 22); 04–06 November 2008;

Hammamet, Tunisia, Abstract Volume. 2008. p. 89

[13] Uwizeye F. Les Roches Mafiques et ultramafiques de Kibungo. Unpublished report. Service Géologique, Ministère des Mines, de l'Industrie et de l'Artisanat, République Rwandaise; 1987. p. 60

[14] Theunissen K, Hanon M, Fernandez-Alonso M. Carte géologique du Rwanda, 1/250000. Service Géologique, Ministère des Mines, de l'Industrie et de l'Artisanat, République Rwandaise; 1991

[15] Dewaele S, Henjes-Kunst F, Melcher F, Sitnikova M, Burgess R, Gerdes A, et al. Late Neoproterozoic overprinting of cassiterite and columbite – Tantalite bearing pegmatites of the Gatumba area, Rwanda (Central Africa). Journal of African Earth Sciences. 2011:1-17

[16] Baudet D, Hanon M, Lemonne E, Theunissen K. Lithostratigraphie du domaine sédimentaire de la chaîne kibarienne au Rwanda. Annales de la Société Géologique de Belgique. 1989; **112**:225-246

[17] Shepherd TJ, Rankin AH, Alderton DHM. A Practical Guide to Fluid Inclusion Studies. Glasgow and New York: Blackie; 1985. p. 239

[18] Möller P, Dulski P, Szacki W, Malow G, Riedel E. Substitution of tin in cassiterite by tantalum, niobium, tungsten, iron and manganese. Geochimica et Cosmochimica Acta. 1988;**52**:1497-1503

[19] Rösler HJ. Lehrbuch der Mineralogie. Leipzig: VEB Deutscher Verlag für Grundstoffindustrie; 1983. p. 833

[20] Beurlen H, Da Silva MRR, Thomas R, Soares DR, Olivier P. Nb-Ta-(Ti-Sn) oxide mineral chemistry as tracer of rare-element granitic pegmatite fractionation in the Borborema Province, northeastern Brazil. Mineralium Deposita. 2008;**43**:207-228

[21] Melcher F, Graupner T, Sitnikova M, Oberthür T, Gäbler HE, Bahr A, et al. Herkunftsnachweis von Columbit-Tantalit-Erzen –Status-quo-Bericht. BGR Open File Report. 2009;**11082/09**

[22] Abdalla HM, Matsueda H, Obeid MA, Takahashi R. Chemistry of cassiterite in rare-metal granitoids and the associated rocks in the Eastern Desert, Egypt. Journal of Mineralogical and Petrological Sciences. 2008;**103**:318-326

[23] Pohl W. Economic Geology, Principles and Practice: Metals, Minerals, Coal and Hydrocarbons — Introduction to Formation and Sustainable Exploitation of Mineral Deposits. Wiley-Blackwell; 2011. p. 663

[24] Linnen RL, Keppler H. Columbite solubility in granitic melts: Consequences for the enrichment and fractionation of Nb and Ta in the Earth's crust. Contributions to Mineralogy and Petrology. 1997;**128**:213-227

[25] MINETAIN. Géologie de Bugarura-Kuluti. Unpublished Report in RMCA Archives-Tervuren. 1965

[26] Bodnar RJ, Vityk MO. Interpretation of microthermometric data for H2O-NaCl fluid inclusions. In: De Vivo B, Frezzotti ML, editors. Fluid Inclusions in Minerals, Methods and Applications. Blacksburg. VA: Virginia Tech; 1994. pp. 117-130

[27] Heinrich CA. Geochemical evolution and hydrothermal mineral deposition in Sn (-W-base metal) and other granite-related ore systems: Some conclusions from Australian examples: Mineralogical

Association of Canada. Short Course Handbook. 1995;**23**:203-220

[28] Wood SA, Samson IM. Solubility of ore minerals and complexation of ore metals in hydrothermal solutions. Reviews in Economic Geology. 1998;**10**: 33-80

[29] Haung HH, Anttila K, Roine A. Eh-pH diagram. In: Outokumpu HSC Chemistry for Windows: Chemical Reaction and Equilibrium Software with Extensive Thermochemical Database. Pori, Finland: Research Oy; 2002

[30] Linnen RL. PIG Comments and Questions, Reply. 2004a. Available from: http://www.minsocam.org/MSA/Special/Pig/Pig_CQ/PIG_CQ/Kjellman.html

[31] Linnen RL. Ferrocolumbite-manganotantalite trends in granites and pegmatites: Experimental and natural constraints. Geological Society of America Progr Abstr. 2004b;**36**

[32] London D, Evensen JM, Fritz E, Icenhower JP, Morgan GB, Wolf MB. Enrichment and accommodation of manganese in granite - pegmatite systems. In: 11th an. Goldschmidt Conference Abstract 3369. 2001. CDR PDF version

[33] Henry DJ, Guidotti CV. Tourmaline as a petrogenetic indicator mineral: An example from staurolite grade metapelites of new Maine. American Mineralogist. 1985;**70**:1-15

[34] Dingwell DB. The structures and properties of fluorine-rich magmas: A review of experimental studies. Canadian Institute of Mining and Metallurgy. 1988;**39**:1-12

[35] London D. Granitic pegmatites. Transactions of the Royal Society of Edinburgh: Earth Sciences. 1996;**87**:305-319

[36] Pollard PJ, Pichavant M, Charoy B. Contrasting evolution of fluorine- and boron-rich tin systems. Mineralium Deposita. 1987;**22**:315-321

[37] Linnen RL. The solubility of Nb-Ta-Zr-Hf-W in granitic melts with Li and Li + F: Constraints for mineralization in rare metal granites and pegmatites. Economic Geology. 1998;**93**:1013-1025

[38] Fetherston JM. Tantalum in Western Australia, geological survey of Western Australia. Mineral Resources Bulletin. 2004;**22**:161

[39] Cerny P, Blevin PL, Cuney M, London D. Granite - related ore deposits. Economic Geology. 2005;**100**:337-370

[40] Burnham CW. Magmas and hydrothermal fluids. In: Barnes HL, editor. Geochemistry of Hydrothermal Ore Deposits. 2nd ed. New York: John Wiley; 1979. pp. 71-136

[41] Taylor JR, Wall VJ. The behavior of tin in granitoids magmas. Economic Geology. 1992;**82**:403-420

[42] Jackson NJ, Halliday AN, Sheppard SMF, Mitchell JG. Hydrothermal activity in the St. just mining district, Cornwall. In: Evans AM, editor. Metallization Associated with Acid Magmatism. Vol. 6. London: Wiley; 1982. pp. 137-179

[43] Günther MA. Flüssigkeitseinschlüsse und geologisches Umfeld zentralafrikanischer Sn-, W- und Au-Lagerstätten (Rwanda und Burundi), Unpublished PhD thesis. Naturwissenschaftlichen Fakultät der technischen Universität Carolo-Wilhelmina zu Braunschweig; 1990. p. 144

[44] Beer KE, Ball TK. Tin and tungsten in pelitic rocks from S.W. England and their behavior in contact zones of granites and in mineralized areas. Proceedings of the Ussher Society. 1986; **6**:330-337

[45] Ngaruye JC. Petrographic and geochemical investigation of Sn - W - Nb - Ta - pegmatites and mineralized quartz veins in southeastern Rwanda. 2011. Avilable from: https://scholar.ufs.ac.za/items/afbdb70f-c486-4c05-b1ab-ff915ca828b4

Chapter 3

Volatile Composition of Fluid Inclusions in Gold-Bearing Quartz Veins Analyzed by Solid-Mass Spectrometry: Method and Contributions to the Orogenic Metallogenic Model and Exploration

Damien Gaboury

Abstract

This chapter provides details about the analytical procedures and examples from published papers of the contributions of this technique to the understanding of, (1) the signification of ethane; (2) the hydrothermal reactions; (3) the water-poor and CO_2-rich fluids; and (4) the source of gold. Gold deposits used as examples range in age from the Archean to the Paleozoic era. Ethane was established as an indicator (proxy) of organic thermal maturation at depth. Hydrothermal reactions inducing graphite precipitation in the host rocks and mineralized structures and consumption of CO_2 and CH_4 were documented in the Otago Schist district in New Zealand, accounting for the low-CO_2 mineralizing fluids, and in the Keraf Zone in Sudan, accounting for the various fluid signatures. Water-poor and CO_2-rich fluids associated with high-grade gold or world-class gold deposits also result from hydrothermal reactions involving the consumption of water and ethane. Gold transport as nanoparticles or as hydrocarbon-gold complexes is related to water-poor, hydrocarbon-rich fluids and associated with high-grade gold mineralization. The sedimentary source of gold is demonstrated by the presence of ethane, which is a tracer of organic matter metamorphism in deep sedimentary rocks at amphibolite facies, providing gold extraction from nodular pyrite during transformation to pyrrhotite.

Keywords: fluid inclusions, ethane, hydrothermal reactions, gold exploration, hydrocarbons, orogenic gold deposits, sedimentary gold source, Otago Schist, Keraf Zone, Abitibi belt, Houndé belt

1. Introduction

For hydrothermal deposits, fluid compositions and temperatures are among the most important parameters for their formation. However, these parameters are not easily established, as direct information only lies in small (<30 μm) preserved fluid inclusions [1], which are the only direct witnesses of fluids involved in the formation of hydrothermal deposits. Numerous techniques have been developed to analyze fluid inclusion volatile compositions individually or in bulk [2–4].

Individual analyses of fluid inclusions rely on their representativeness to the mineralizing event. Individual analyses include the non-destructive Raman and destructive LA-ICP-MS techniques. Raman is limited by uncertainty, which is composition and density dependent [5]. For LA-ICP-MS, it is difficult to quantify *in situ* the composition of a single fluid inclusion due to the short signal, the low volume of material extracted, and the laser-induced matrix effect [6].

Bulk analyses include fast heating or mechanical crushing of fluid inclusions and gas analyses mainly by mass spectrometry and gas chromatography (e.g., [2, 7–14]). These bulk techniques give a composite fluid composition from all fluid inclusions in a sample. Comparatively, the solid-sonde approach yields real-time, temperature-resolved mass spectrometric analyses of gases released directly under vacuum by fluid inclusion decrepitation during controlled heating, hence providing more confidence about fluid compositions.

This chapter aims to present improvements to a solid-probe mass spectrometry analytical technique first described by Guha et al. [15] and case studies that contributed significantly to the metallogenic model for orogenic gold deposits, and implications for gold exploration. Finally, this chapter aims to provide information to other laboratories for developing similar apparatus and key parameters for analyzing and interpreting data.

2. Apparatus description

The first description of the apparatus was provided by Gaboury et al. [16] at the time it was under development. The final configuration is more complex and relies on a different mass spectrometer. The system is composed of seven main components (**Figure 1A,B**): (1) a high vacuum system; (2) a heating and sample holder probe; (3) a high-pressure cooling water pump; (4) a heating control module; (5) a quadrupole mass spectrometer (QMS); (6) a backing pump for sample insertion; and (7) a vacuum gauge. The vacuum pumping system includes a Pfeiffer Vacuum turbomolecular pump and an oil-free backing pump to avoid contamination by hydrocarbons. The final pressure is in the range of 1 E^{-8} mbar. The solid-probe was fabricated by GV Instruments and consists of a removable stainless-steel tip holder coupled with an electric heater and a thermocouple (**Figure 1C**). The water pump provides cooling of the probe to prevent thermal dilation and atmospheric contamination of the vacuum. Water flow is kept at a constant rate to limit temperature variation at the probe tip. A crushed solid mineral sample is inserted into a silica tube, about 10 mm long by 1 mm in diameter (**Figure 1C**). This silica tube is held by the stainless-steel tip. Further technical detail about the probe is given in Guha et al. [15]. Probe heating is controlled by an Omega programmable CN8200 temperature controller, driven by Omega proprietary software. The rate of heating of the probe is fully programmable up to a maximal temperature of 600°C. The maximum rate of heating is about 50°C per second,

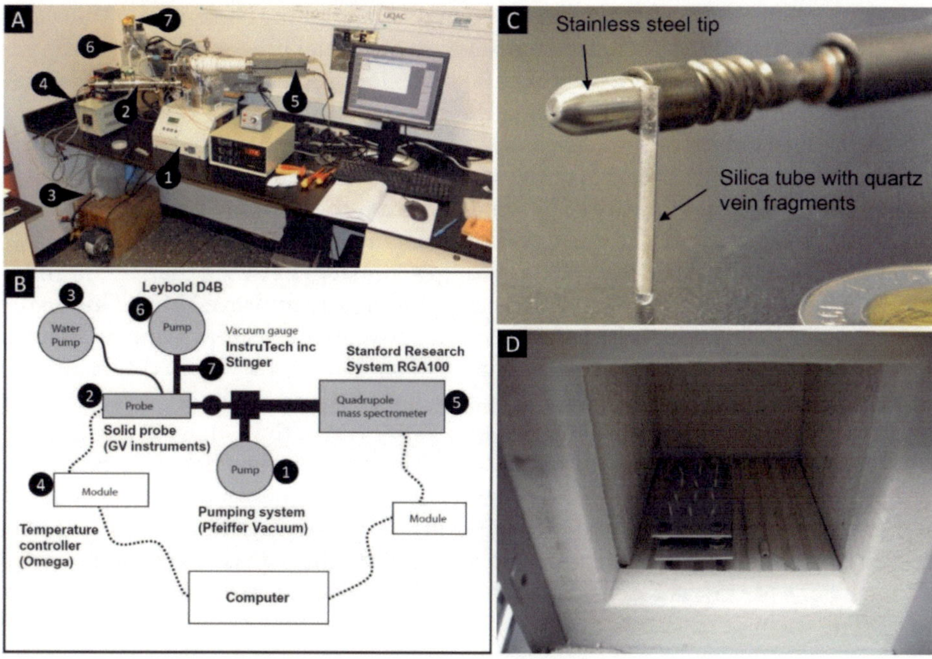

Figure 1.
System characteristics. (A) The analytical system as installed at the University of Québec at Chicoutimi with main components numbered. (B) Schematic representation of the system with the main components numbered, modified from Gaboury et al. [16]. (C) Silica tube (1 cm long) with the stainless-steel solid-probe hold tip. (D) Quartz sample in silica tubes in the oven for a night at 100°C before analysis for removing absorbed water.

whereas the slowest is 0.1°C per minute. Temperature variation relative to a point set by the controller is better than 0.6°C (average of 0.2°C) over a range of 100–500°C. The mass spectrometer is a quadrupole Stanford Research System RGA100 with an open ion source operated at 70 volts with a secondary multi-channel continuous-dynode electron multiplier (CDEM) operated at 1000 volts. Mass resolution is better than 0.5 AMU over the range of 1–100 AMU. Detection limits with the CDEM are in the range of 10 ppb or a gas partial pressure of 1 E^{-13} mbar. RGA software version 3.1 was used to operate the spectrometer and collect data. Finally, a Leybold D4B mechanical pump coupled with a vapor oil filter is used to purge the solid-probe at pressures <6 mbar before sample insertion in the high vacuum, preventing the QMS from atmospheric contact and contamination. This operation is guided by a convectional enhanced Pirani vacuum gauge module (CVM-211 Stinger by InstruTec).

2.1 Principles of fluid release and analysis

A sample of pure mineral hosting fluid inclusions is inserted at the end of the solid-probe tip and heated under vacuum. As the sample is heated, the fluid inclusions decrepitate due to internal overpressure. Decrepitation can release gases from a single inclusion or a group of inclusions [10]. The released volatiles are directly analyzed by the QMS in terms of AMU abundance, measured as the partial pressure of N_2 equivalent. The QMS analyses are performed in dynamic mode (i.e., during continuous vacuum pumping), and as a result, gases released from decrepitation are analyzed and then pumped out. Expelled gases from fluid inclusion decrepitation are expressed by an increase of the partial pressure relative to the background value during real-time monitoring. A continuous increase of partial pressure for a specific gas during heating indicates that bursts of fluid inclusions increased the internal partial pressure at a rate greater than pumping. The return to background values after the end of inclusion bursts occurs in the range of ~10 seconds. Analyzing in a dynamic state implies that the heating rate of the sample, vacuum pumping rate, scanning speed by the QMS, and quantity of gas released from the sample (sample size) have to be matched for optimal measurements. These parameters are addressed below.

Because there is no interface (no inlet for gas transfer) between the sample and mass spectrometer, the precision of the analysis is greatly improved [15]. This feature enables a direct correlation without correction between the decrepitating temperature and the volatile composition of the fluid inclusions. Therefore, binary diagrams of gas pressure (abundance) versus decrepitating temperature can be constructed as the first data representation. Within the range of 1–100 AMU, the following gases can be analyzed: CO_2, N_2, CO, CH_4, SO_2, H_2S, H_2O, He, Ar, and some light hydrocarbons of the type C_nH_{2n+2}, such as C_2H_6, C_3H_8, and C_4H_{10}. All of these gases are considered major to rare components of fluid inclusions [4]. Distinction of N_2 and CO cannot be achieved based solely on their masses. However, given that CO is a rare component of fluid inclusions [4, 15], it is considered that 28 AMU mainly represent the N_2 component of fluid inclusions.

2.2 Sample selection

Between 5 and 10 mg of hydrothermal minerals are required for the analysis. Ideally, a pure mineral is preferred because the release of volatiles is from one material with specific mechanical properties. This parameter is fundamental as inclusion decrepitations are influenced by various factors, notably composition (liquid and gas proportions), trapping temperature (homogenization temperature), size and shape, and most importantly, the mechanical resistance of the host material relative to internal pressure exerted by the heated fluid inclusions [17, 18].

Quartz gives excellent results for inclusion decrepitation as the material is brittle, even at high temperatures (500°C). Carbonates yield acceptable results, though they become more ductile at high temperatures. Carbonates also contribute to the CO_2 signal by thermal decomposition at high temperatures (>425°C: [10]). Pyrite yields good results, but thermal releases of H_2S and SO_2 occur at >400°C as indicated by the spotty melting on the grain surfaces. In short, quartz, which is one of the most common minerals in hydrothermal deposits, is also the best material for fluid inclusion analyses by thermal decrepitations.

2.3 Sample preparation

Selected parts of pure quartz grains are chosen and then coarsely crushed. Pure quartz fragments between 1 and 2 mm are hand-picked under a binocular microscope, dry crushed in an agate mortar, and then placed in the silica tube. Hand crushing is a delicate stage, as grain reduction must be done without destroying fluid inclusions by excessive grinding pressure. Numerous tests have shown that grain sizes are not critical as long as quartz particles cover the range of 0.01–0.5 mm. A large distribution of particle sizes (0.01–0.5 mm) is preferred over sieved samples as it provides more surface exposures for fluid inclusions of various sizes. Samples inserted in a silica tube are heated in a small oven overnight at 100°C before analysis to eliminate absorbed water (**Figure 1D**). All the sample preparation and loading in the apparatus are done with nitrile gloves and without the use of alcohol or organic solvent to prevent contamination.

2.4 Sample analysis procedures

Each sample is taken directly from the oven and inserted in the probe tip using fine stainless-steel tweezers. The probe is inserted into the apparatus and pumped by the roughing pump. When the rough vacuum is reached (~6 mbar), a valve is opened to connect the sample with the high vacuum section of the QMS. Pumping down to the optimal pressure (~1 E^{-8} mbar) for QMS analysis takes approximately 1 hour. When the optimal pressure is reached, the sample is linearly heated from 30 to 500°C. During heating, volatile spectra are collected by the QMS in real time, and temperature is also synchronously recorded by the Omega software. One sample analysis takes about 2 1/2 hours from the sample insertion to the end of analysis.

2.5 Heating rate and QMS analytical cycle

The release of sufficient gases for the analysis is directly related to the sample size (density of fluid inclusions) and the heating rate [16]. If the heating rate is too slow, each burst is monitored individually, and the quantities of gases released can be too low to be analyzed by the instrument. If the heating rate is too fast, only a bulk fluid composition is obtained. Optimal analytical performance is also dependent on the scanning speed of the QMS. If the speed is too fast, the analytical precision is weaker; if the speed is too slow, released gases can be pumped out before being scanned by the QMS. Since the size of the sample is physically limited by the solid-probe configuration, only the heating rate of the sample and the scan speed of the QMS are configurable.

Numerous tests were performed to determine the best heating rate using different quartz samples from various types of gold deposits, and a heating rate of 6°C per minute was selected. For most samples, this rate is fast enough to provide sufficient released gases to produce a QMS analytical peak above the detection limit (3σ of background) without QMS saturation. To accommodate this heating rate, the QMS scanning cycle was set at a 5 second(s) speed. Each volatile species is then analyzed during 0.5 s counting time, providing a complete volatile analysis every 0.5°C heating increment. With this configuration, a sample analysis is composed of more than 800 volatile analyses over a heating range of 100–500°C. This setup provided the most accurate and stable analyses of each gas species over the full temperature range.

2.6 Gases and mass selection for monitoring

With the QMS, up to 10 masses can be monitored continuously. The volatiles H_2O, CO_2, N_2, C_2H_6, He, Ar, CH_4, SO_2, H_2S, and H_2 were selected as they are the most common species analyzed from fluid inclusions (e.g., [19]) or the most interesting species (N_2, Ar, He) for fluid source interpretation (e.g., [20]).

During ionization in the QMS, gas molecules fragment into lighter daughter molecules [21]. For example, CO_2 (44 AMU) breaks up into lighter molecules of 28, 16, and 12 AMU, corresponding to CO-N_2, O, and C masses, respectively. Consequently, for most volatiles, there is a percentage expressed by their parental, and to a lesser proportion, by their daughter molecules. The fragmentation proportions are specific to ionization parameters and mass spectrometer configuration [22]. For the Stanford Research System RGA100, the fragmentation processes and proportions of parental and daughter molecules for each gas species are well established and provided by the manufacturer. Therefore, gases can be monitored using either the parental or the daughter molecule masses, as long as a correction factor is applied to calculate their absolute abundance. **Table 1** lists the AMU masses used to monitor the selected gas species. Selection of masses was based on three parameters: (1) avoiding mass interference; (2) providing a better linear signal for the background profile (see below); and (3) use of less abundant daughter molecule mass for analyzing a specific gas species by CDEM for improved signal/noise ratios.

2.7 Background profiles

One of the major issues for quantitative estimation and percentage mole calculation of volatile content is the establishment of linear background gas profiles over the heating range. As stated by the "ideal gas law," partial gas pressure increases with temperature, hence inducing sloped background profiles for most of the monitored gases. To overcome this phenomenon, the valve and short tube between the solid-probe and the mass spectrometer are heated (>100°C) at a constant temperature. Consequently, the background signal is increased for some gases, but more importantly, it is flattened for most gases. Alternatively, some molecular masses appear more sensitive to heating than others, even if the mass spectrometer software includes an internal algorithm for temperature correction. As a consequence, some gas pressures were monitored using their second or third ionization masses of daughter molecules (**Table 1**) as they proved to generate a flatter background profile. The background profile for each volatile is based on a compilation from numerous background analyses.

Gas	AMU	Prop[1]	Fragment	CDEM[2]	Cor.[3]
H_2O	18	74.40%	parental	off	1.34
CO_2	44	78.43%	parental	off	1.28
N_2	28	92.59%	parental	off	1.08
H_2	2	95.23%	parental	off	1.05
Ar	40	90.50%	parental	off	1.10
He	4	100.0%	parental	on	1
CH_4	14	7.35%	daughter	on	4
C_2H_6	27	15.06%	daughter	on	3.5
H_2S	33	21.27%	daughter	on	4.7
SO_2	64	57.08%	parental	on	1.75

1. Proportion of the AMU as provided by the QMS manufacturer
2. Multi-channel continuous-dynode electron multiplier
3. Factor of correction applied to the AMU abundance.

Table 1.
Parameters of monitored volatiles.

2.8 Correction for molecule ionization and mass interferences

Since ionization induces molecule fragmentation, a correction factor should be applied to the reduced data in order to correct for the relative abundance of the parental and daughter molecule masses used for gas monitoring. These correction values are listed in **Table 1**. For all gases measured with their parental molecules, the correction was based on the proportion of parental molecule masses. For all other gases, such as CH_4, C_2H_6, and H_2S, measured with daughter molecule masses, the correction factor was established based on the inversion matrix calculation function of the QMS software. Additionally, some mass interferences should be corrected. As N_2 is measured at 28 AMU, CO_2 and C_2H_6 fragment to daughter molecules of the same mass in proportions of 8.62 and 47.28%, respectively, and the proportional contribution of those gases has to be removed to obtain real N_2 values. Finally, as N_2 cracks to daughter molecules of mass 14 AMU at a portion of 6.66%, CH_4 measured at 14 AMU must also be corrected for the potential contribution of N_2.

2.9 Data reduction and mole percentage

A simple graph of partial pressure (mbar) of volatiles relative to the temperature (°C) is used at first to visualize the relative abundance of volatiles from fluid inclusions decrepitating over the heating range (**Figure 2A**). Since fluid inclusion decrepitations cause small pressure bursts that look like a sawtooth signal above the

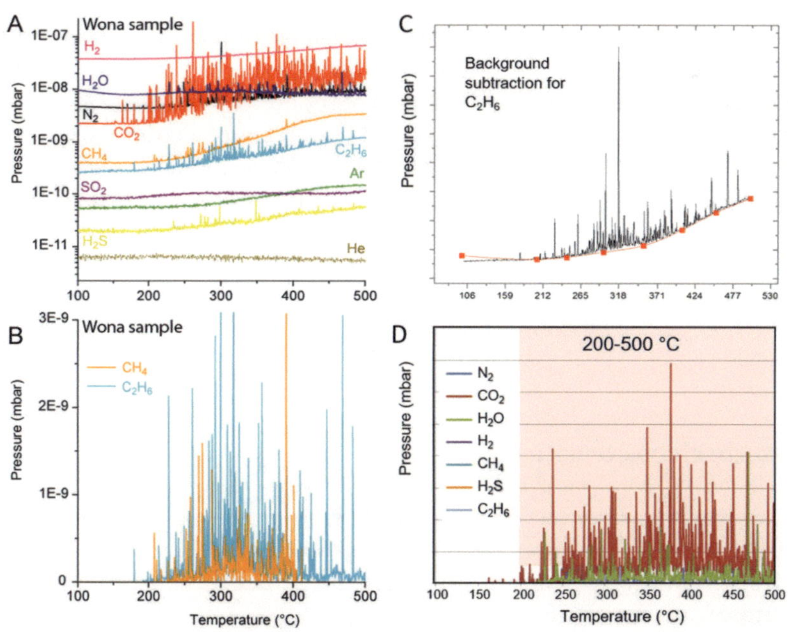

Figure 2.
Mass spectrometric time-resolved signal generated over the temperature range of 100–500°C for the 10 volatiles analyzed. The sawtooth signal indicates fluid inclusion decrepitations above the background. For this sample, CO_2, H_2O, C_2H_6, CH_4, and H_2S were released. (B) Example of treated signal for C_2H_6 and CH_4 with background removed. (C) Example of background subtraction with the OriginPro-8 software. (D) Example of the residual signal with the background removed for all the volatiles. The mole% is calculated over a temperature range of 200–500°C to cover the signal range. (A-B) from Gaboury [23]; used with permission of the Geological Society of America".

background, any contamination from air and other substances, such as a synthetic dust fiber, can be easily detected (**Figure 2A**). The raw data are converted to quantitative values by background signal subtraction using the OriginPro-8 software (**Figure 2B**). This software has an integrated function for fitting background reference profiles and subtraction (**Figure 2C**). With the background removed, the total pressure for each volatile is calculated by the summation of all the partial pressure analyzed over the temperature range generally from 200 to 500°C (**Figure 2D**). For comparison between samples, volatile pressure data are converted to mole percent (mole%). As the fluid inclusions in natural samples are not uniformly distributed, all samples were analyzed at least twice to ensure the representativeness of the released gas signal.

3. Fluid inclusion characteristics

Multiple events are recorded in structurally controlled quartz-vein-type gold deposits, particularly in orogenic gold deposits. These events include quartz precipitation and repeated fracturing [24–26]; they also include post-mineralization fluid overprint [27]. Over time, fluid inclusions can also undergo physical modification expressed as necked inclusions (e.g., [28]). When studying fluid inclusions, it is thus fundamental to determine their genetic link with gold (e.g., [29, 30]). Instead of attempting to link individual fluid inclusions to the gold event—which is probably not feasible in most situations—the bulk gas analysis gives an average percentage of each gas among all the fluid inclusions.

The petrographic documentation is also needed to support the interpretation that gas is released from fluid inclusion assemblages expected from vein-type gold deposits. Phase assemblages were established at room temperature (22°C). As would be expected for vein-type gold deposits, quartz samples are characterized by assemblages of fluid inclusions that are mostly distributed as secondary origin clusters and trails (**Figure 3**). Fluid inclusions come in a wide variety of forms and colors, from translucent to black and ovoidal to flame-shaped (**Figure 3**). In orogenic gold deposits, it is usual to distinguish between four types. **Figure 3** shows the four different types: (1) ovoid, 1 phase liquid or gaseous; (2) ovoid, 2 phases liquid with a small bubble of darker gas; (3) ovoid, 1 phase dark gaseous; and (4) flame-shaped, 1 phase dark gaseous.

4. Case study characteristics

The case studies presented below are only from orogenic gold deposits. Because gold is commonly associated with secondary fluid inclusions [24, 26], a bulk analysis technique is appropriate to get the global composition of the mineralizing fluids.

Although orogens cover the entire geological age spectrum [31], orogenic gold deposits are distributed in three main periods [32, 33]: (1) Neoarchean (2.75–2.55 Ga); (2) Paleoproterozoic (2.1–1.75 Ga); and (3) Neoproterozoic-Phanerozoic (<800 Ma). The case studies are detailed below, according to the three main geological periods. All the samples were taken by the author during field missions, hence ensuring the representativeness of the material analyzed.

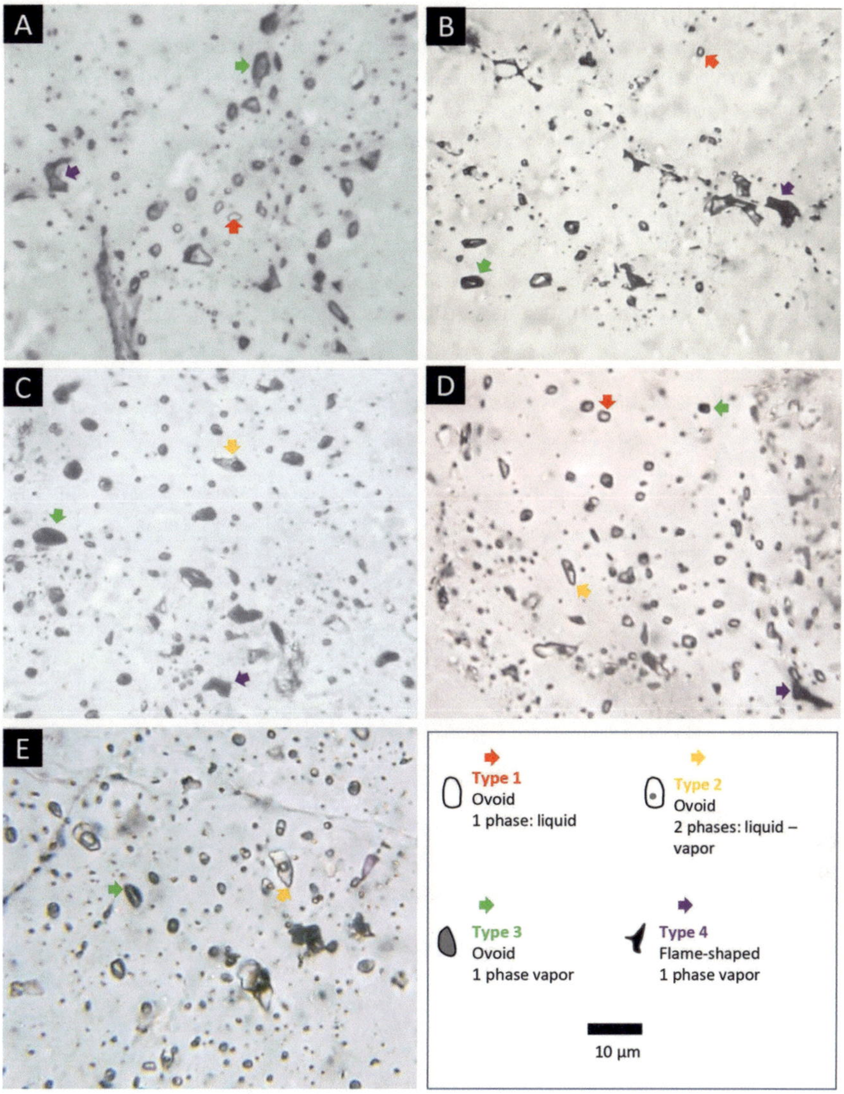

Figure 3.
Photomicrographs showing the petrographic characteristics used for defining the four types of fluid inclusions. The arrow highlights the various types. (A) Detour Gold, sample DG-43. (B) The High-Grade Zone (HGZ) of the Perron deposit, sample RC-13. (C) Mana district, Siou deposit, sample 513B). (D) Keraf district, sample Sherick-118-3. (E) Otago Schist, sample MT-8A.

4.1 Neoarchean Abitibi belt, Canada

4.1.1 Detour gold deposit

The Detour Gold deposit is the largest producing gold mine in Canada with a large gold reserve of 20 Moz (Agnico-Eagle Mines). It is located in the northern Archean Abitibi belt. A swarm of quartz and sulfide veinlets defines a subvertical 200–350 m-wide East-trending corridor hosted by verticalized mafic to ultramafic lava flows and hyaloclastic rocks, metamorphized to upper greenschist facies [34]. A total of 28 samples were analyzed from drill cores taken along three sections of the deposit [23]. Volatile compositions of fluid inclusions are presented in **Figure 4A**. Fluids are composed in variable proportions of H_2O, CO_2, N_2, and CH_4. Ethane is present in four samples. Some samples recorded water-poor and CO_2-rich fluids.

4.1.2 Perron gold deposit

The Perron deposit is an active-exploration gold project located in the northern part of the Abitibi belt and is currently under extensive drilling. This project bears

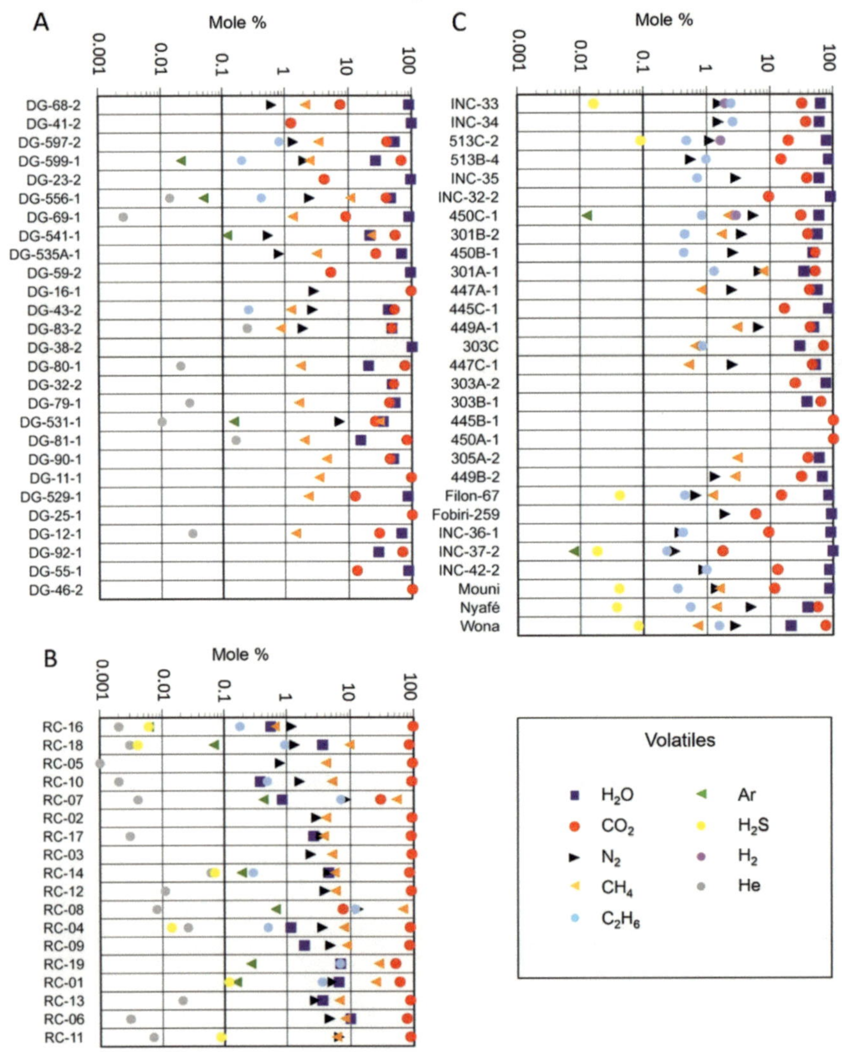

Figure 4.
Logarithmic graphs of mole% of volatile released from fluid inclusions. Samples are ordered from left to right according to the released total pressure (mbar)—quantity of volatiles released by fluid inclusion decrepitation. (A) Detour Gold, data from Gaboury [23]. (B) The HGZ of the Perron deposit, data from Gaboury et al. [35]. (C) Mana district, data from Gaboury [23].

some of the highest gold-grade mineralization for Archean vein-type deposits worldwide [35]. The High-Grade Zone (HGZ) is a steeply north-dipping orogenic ore shoot. It is traced by core drilling over 500 m laterally, from the surface down to >1260 m vertically with a 75° eastern plunge. The true thickness ranges from 1 to 4 m. The HGZ is composed of massive whitish-to-grayish quartz veins with traces of iron carbonates, green chlorite, and tourmaline. Sphalerite is the most abundant sulfide, reaching >5% locally, as is pyrite. Visible gold grains are abundant and associated with sphalerite, chlorite ribbons, and freely in quartz. The HGZ is developed at the interface of a narrow (1–5 m) deformed mafic dyke cutting a metamorphized rhyolite. A total of 18 samples were analyzed from the HGZ to cover its vertical extend and various gold grades [35]. Fluid compositions (**Figure 4B**) are similar to fossil gas. They are water-poor but rich in CO_2 and composed of hydrocarbons (methane and ethane and possibly butane and propane and other unidentified organic compounds), with N_2 and traces of Ar, H_2S, and He.

4.2 Paleoproterozoic Houndé belt in Burkina Faso

The Mana district (~10 Moz: Endeavor Mining Inc.), in the Paleoproterozoic Houndé belt of Burkina Faso, hosts at least three different mineralization styles [36]: (1) high-grade (5–10 g/t) quartz-carbonate veins (Filon-67 and Siou); (2) low-grade and large volume of sulfurized greywacke (Yaho: ~1 g/t); and (3) intermediate grade (~2.5 g/t) in strongly deformed and silicified rocks (Wona mine). The deposits are scattered within a ~ 100 km^2 area, and are hosted in various rock types at greenschist facies, such as strongly deformed and verticalized sedimentary and volcaniclastic rocks interbedded with mafic sills and lava flows. Graphitic and pyritic shale horizons are common in the sedimentary rocks. Igneous rocks of various compositions intrude the strata. A total of 29 samples were analyzed from various gold deposits, sampled from drill cores [23]. Fluids are composed of H_2O, CO_2, N_2, CH_4, C_2H_6, and ± H_2S in variable proportions (**Figure 4C**). Some samples recorded water-poor and CO_2-rich fluids.

4.3 Neoproterozoic Keraf Suture zone in Sudan

The Neoproterozoic Nubian Shield in northern Sudan is an important gold production area. The Keraf Suture zone is the most prominent feature for gold deposits. The ~500 km-long and ~ 50–100 km-wide N-trending continental arc suture [37–39] occurs between the Neoproterozoic Nubian Shield in the east and the older East Sahara Ghost Craton to the west. The Keraf Suture formed during the Neoproterozoic consolidation of Gondwana. Gold deposits are located within and along the Keraf Suture zone. About 59 samples of gold-bearing quartz veins were analyzed (**Figure 5A**). Samples are from more than 20 different sites scattered in a large area of 250 km (NW-SE) by 100 km (SW–NE) in northern Sudan [40]. In the Gabgaba area, WG-03, UTM, and Central are in commercial production. Shereik is an important deposit with mineral resources. Other deposits and sites are more marginal, with only artisanal gold production.

From fluid inclusions, six types of fluids are distinguished based on specific volatile contents and proportions [40]. Type 1 is an aqueous-carbonic fluid, dominated by water, with a complex signature including H_2O, CO_2, N_2, CH_4, C_2H_6, and ± H_2. Type 2 fluids are comparable to type 1 with ethane content but with less water (< 50 mole%). These fluids have a complex signature with CO_2, N_2, CH_4, C_2H_6, ±H_2, Ar, and H_2S. Type 3 is an aqueous-carbonic fluid similar to type 1, dominated by water but without ethane. Type 4 is an aqueous-carbonic fluid without hydrocarbons but containing nitrogen and some hydrogen. Type 5 are fluids devoid of water, with variable proportions of CO_2 (0–100 mole%), N_2 (0–62.8 mole%) and CH_4 (0–42.4 mole%). Finally, type 6 is composed only of water. The largest gold deposits recorded fluids with ethane (types 1 and 2) and also type 5, which is a water-poor and CO_2-rich fluid.

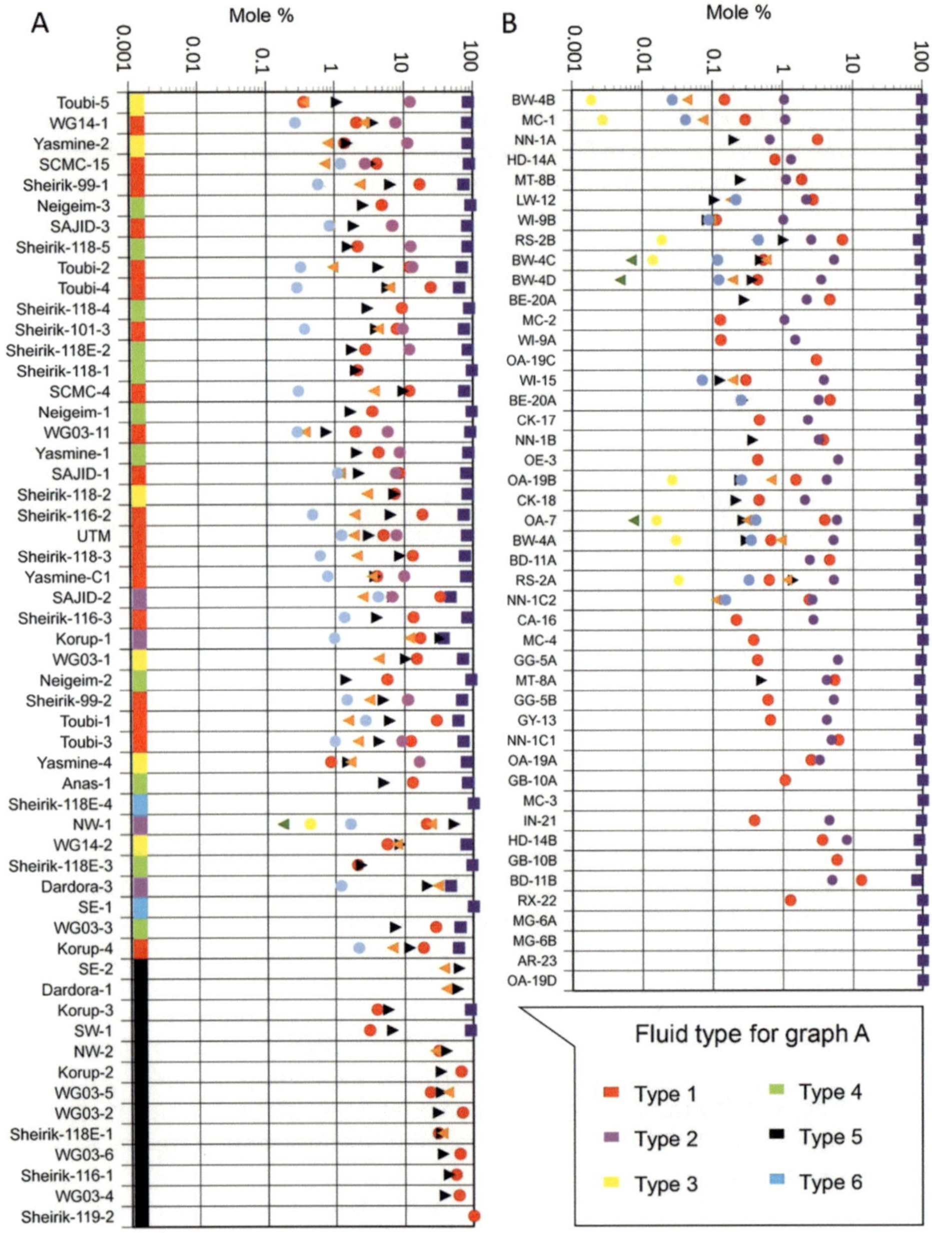

Figure 5.
*Logarithmic graphs of mole% of volatile released from fluid inclusions. Samples are ordered from left to right according to the released total pressure (mbar)—quantity of volatiles released by fluid inclusion decrepitation. (A) Keraf district, data from Gaboury et al. [40]. (B) Otago Schist district, data from Gaboury et al. [41]. Symbols for volatile species as in **Figure 4**.*

4.4 Mesozoic Otago Schist belt in New Zealand

The Mesozoic Otago Schist belt of southern New Zealand is a ~ 100 km-wide Mesozoic accretionary complex composed mostly of greywacke and argillite with minor basalts [42, 43]. The belt defines a broad antiformal structure exposing different metamorphic facies. In the core, rocks are at upper greenschist facies (biotite-garnet zone). Conversely, rocks in the NE and SW flanks are at lower greenschist to sub-greenschist facies [44]. This belt is well known for its uniform host rock lithology, geochemistry, and mineralogy over 40,000 km^2 of exposure [45]. This belt is also devoid of igneous intrusions associated with either metamorphism or post-metamorphic gold mineralization [46]. Over 200 gold-bearing quartz veins are hosted in Otago Schist [47, 48]. The subvertical quartz veins commonly strike northwest and cut the shallow dipping schistosity at high angles. Some shallow dipping shear zones are also gold-bearing, such as the world-class Macraes mine [49, 50], and the Rise and Shine deposit [51]. A total of 45 samples from 19 gold deposits were used for volatile analyses [41]. In comparison with other orogenic gold deposits worldwide (**Figure 5B**), the gold-bearing fluids are dominantly aqueous (H_2O > 82.13 mole%) with a very low concentration of CO_2 (< 12.93 mole%), but with significant contents of H_2 (up to ~8 mole %) and C_2H_6 (up to ~0.5 mole%).

5. Contributions to the orogenic model for gold deposits

The scientific contributions from the results generated by solid-probe analyses of fluid inclusions from the studied cases are described below.

5.1 Ethane signification

Ethane (C_2H_6) is present in fluid inclusions from orogenic gold deposits ranging in age from ~2800–~100 Ma ([23]; **Figures 4** and 5). Methane and ethane are the most common light hydrocarbons forming organic compounds called alkanes with a general formula of C_nH_{2n+2}. Methane is a common volatile in fluid inclusions of orogenetic gold deposits, but ethane is rarer [52–54]. Ethane is more difficult to analyze because of its lower concentrations.

Three processes can generate alkanes (e.g., [55]): (1) bacteriological activity; (2) thermogenic maturation of organic matter; and (3) Fischer-Tropsch synthesis. The first two are biogenic, related to organic-rich matter trapped in sedimentary rocks, and alkane generation occurs at low and higher temperatures, respectively. The origin of the light hydrocarbons is typically ascertained using the value of CH_4/{C_2H_6 + C_3H_8}, which is expressed as C_1/C_{2+} [56]. A value of less than 100 indicates the production of thermogenic alkanes. Conversely, C_1/C_{2+} greater than 1000 imply a very low C_{2+} hydrocarbon production [56], which is typically associated with microbial methanogenesis [55]. When CH_4 and C_2H_6 are present in fluid inclusions for the Detour Gold and Mana districts, the C_1/C_2 ratios range from 2.6–25.5 [23], indicative of a thermogenic origin.

By reacting CO and H_2 with catalysts, abiogenic Fischer-Tropsch synthesis can create CH_4 and other organic compounds. Despite arguments that such an abiotic process was occurring in seafloor volcanic hydrothermal settings (e.g., [57]), Schrenk et al. [58] and McCollom [59] concluded that except for methane, production of abiogenic higher alkane remains speculative. In short, C_2H_6 cannot be produced from abiotic hydrothermal reactions. As a result, Gaboury [23] concluded that ethane was a trustworthy indicator of the involvement of carbonaceous and pyritic shales at depth for the formation of gold deposits (see below).

5.2 Hydrothermal reactions

Gold mineralizations in the Otago Schist are young and provide a unique laboratory to test hydrothermal fluid reactions as the belt is considered homogeneous. At the belt scale, the fluid composition is exceptionally homogeneous but particularly with low CO_2 and high H_2 contents (**Figure 5B**). Gaboury et al. [41] proposed that this unusual composition was generated by chemical reactions in the fluids inducing graphite precipitation in the host rocks and gold-bearing structures and consumption of CO_2 and CH_4. The following hydrothermal reactions are considered to account for the final mineralizing fluid composition [40, 41]. Methane can produce H_2 from the following reactions:

$$CH_4 + H_2O = CO + 3H_2 \tag{1}$$

$$CH_4 = C + 2H_2 \tag{2}$$

However, this reaction can be countered by the reduction of CO_2 to CH_4, with associated consumption of H_2, following the reaction:

$$CO_2 + 4H_2 = CH_4 + 2H_2O \tag{3}$$

Ethane can be degraded in hydrothermal fluids by the following reactions, some of which also produce H_2 [23]:

$$C_2H_6 + 2H_2 = 2CH_4 \tag{4}$$

$$C_2H_6 + 4H_2O = 2CO_2 + 7H_2 \tag{5}$$

$$C_2H_6 + 2H_2O = CO_2 + CH_4 + 3H_2 \tag{6}$$

From all these reactions, Gaboury et al. [41] proposed that C_2H_6 was degraded to produce H_2, CO_2, and CH_4 (Eqs. 4–6). Concomitantly, CO_2 was reduced to CH_4 by consumption of H_2 (Eq. 3). Ultimately, H_2 and graphite were produced from eq. 2 by decarbonization of CH_4. The unique homogenous composition of fluids through time and space for the Otago Schist is thus related to the combination of (1) the relatively homogeneous sedimentary rock source for fluid generation by metamorphic dehydration; (2) large-scale hydrothermal chemical reactions for consuming carbon-bearing volatiles; and (3) lack of meteoric or magmatic inputs for fluid mixing.

These hydrothermal reactions active on primitive fluids of metamorphic origin, containing H_2O, CO_2, N_2, CH_4, C_2H_6, and ± H_2 (type 1; **Figure 5A**), were also invoked by Gaboury et al. [40] to account for the six different types of fluids established for the gold deposits in the Neoproterozoic belt in Sudan.

5.3 Water-poor and CO_2-rich fluids

Some world-class gold districts and deposits have CO_2-rich and H_2O-poor fluid inclusions, such as those at the Red Lake [60, 61], Ashanti and Tarkwa goldfields [62–64], and the Detour Gold and Wona deposits [23].

N_2, C_2H_6, and CH_4 are also present in the mineralizing fluids for these gold deposits [65]. There is ongoing discussion over the origins of these peculiar fluids [23, 66]. According to Goldfarb et al. [67], a range of carbon-bearing molecular components within metamorphic C—O—H—S fluids bearing gold could result from the devolatilization of abundant carbonaceous schists and cherts in the Ashanti gold belt. Lüders et al. [68] confirmed a sedimentary source for the generation of the metamorphic fluids in the Ashanti gold belt using the stable carbon isotopic signature in CO_2-rich fluid inclusions from quartz-hosted gold mineralizations. Consequently, it is likely that the metamorphism of sedimentary rocks rich in organic material is the source of these peculiar fluids. Nonetheless, Gaboury [65] established that these fluids are linked to very high-grade or very large gold deposits, suggesting that CO_2-rich and H_2O-poor fluids have untapped potential for producing extraordinary gold deposits.

CO_2-rich and H_2O-poor fluids can be explained by hydrothermal reactions involving the consumption of water and ethane, the latter being the best indicator of gold-bearing carbonaceous materials at depth [23].

The hydrothermal reactions (Eqs. 5 and 6) will consume ethane and water for producing CO_2. Ethane and methane are also common constituents of these deposits (**Figures 4** and 5). In addition, the presence of H_2 in the mineralizing fluids in the Otago Schist (**Figure 5B**) confirms that the chemical reactions (Eqs. 5 and 6) are viable for the ultimate production of CO_2-rich and H_2O-poor fluids.

5.4 Hydrocarbon-rich mineralizing fluids

Gold grades in the HGZ of the Perron deposit are coherently distributed over a steeply plunging >1.2 km-long ore shoot, with a range of 30–300 ppm, 10–50 times greater than usual values for orogenic gold deposits [35, 69]. The gold mineralization was formed at ~370°C, significantly lower and later than the ~600°C metamorphic peak of the host rocks [35]. The calculated conditions of the sulfur fugacity (−6), oxygen fugacity (−28), and pH (~7) were in the optimal window for solubilizing gold with aqueous HS^- complexes at 350°C [35]. However, fluid inclusions are devoid of H_2O (**Figure 4B**), hence precluding an aqueous transport of gold. Fluid inclusions are rich in hydrocarbon-bearing volatiles (CH_4, C_2H_6), with variable concentrations of CO_2, N_2, noble gases (N_2, He, and Ar), and H_2S (**Figure 4B**). The mineralizing fluids included unidentified organic compounds as well as other alkanes, such as butane and, possibily, propane. These fluids are comparable to fossil gas from producing fields, albeit with a higher CO_2 content.

Hydrocarbons have recently been linked to the formation of gold deposits, particularly in sedimentary settings, such as epithermal [70], Carlin-type [71], orogenic [72], and even the giant Witwatersrand goldfields [73, 74]. Petroleum-phase gold transport is now considered a potentially efficient ore-forming process based on experimental data [75, 76], but specific Au-organic/Au-hydrocarbon complexes remain to be identified. According to a recent study [77], in the hydrocarbon-rich ore-forming fluids, gold can react with acetate to generate Au-organic complexes like $Au(CH_3COO)$ and $Au(CH_3COO)_2^-$. In addition, gold can also be transported as colloidal nanoparticles in hydrocarbon-rich fluids [78–80].

On these bases, Gaboury et al. [35] concluded that gold was transported as hydrocarbon-phase metals or as gold colloidal nanoparticles, hence accounting for the gold richness of the HGZ. Recently, a detailed LA-ICP-MS study of pyrites concluded that gold was in part transported by Bi-Te-melts in the HGZ to account for the high concentration of Bi and Te and their correlation with gold [69]. The gold transport by Bi-Te melts is also possible in non-aqueous fluids. In short, hydrocarbon-rich mineralizing fluids at Perron are an extreme case of water-poor and CO_2-rich fluids, with a demonstrated higher capacity for transporting gold and forming high-grade or world-class deposits.

5.5 Sedimentary source of gold

The sources of gold for orogenic deposits have been reviewed by numerous authors (e.g., [29, 81, 82]). Gold can be sourced from intrusion degassing (e.g., [83]) and oceanic basalt devolatilization (e.g., [84–86]). Nonetheless, one of the most significant sources is thought to be carbonaceous and pyrite-rich sedimentary rocks, also known as black shales (e.g., [40, 41, 65, 82, 87–92]).

The sedimentary source involves the expulsion of gold and trace metals from primary pyrites by transformation to pyrrhotite during prograde metamorphism. This model was proposed by Thomas et al. [93] and Large et al. [87] on the basis that the pyrites in the gold veins have trace metal signatures similar to nodular pyrites in the host black shales. Other studies have confirmed this link for the Paleoproterozoic Mana district in Burkina Faso [36] and the Phanerozoic Otago Schist district in New Zealand [88].

The transformation of pyrite into pyrrhotite is also accompanied by the release of sulfur [94, 95], providing the ligand (HS^-) for transporting gold in solution [96, 97]. Thermodynamic modeling has shown that black shales release the maximum amount of free sulfur for linking gold [95]. Other studies have shown that metamorphic fluids are generated at the transition of greenschist and amphibolite facies [98], i.e., at similar temperatures as the transformation of pyrite into pyrrhotite. On these bases, Gaboury [23] concluded that black shale constitutes the best source for forming orogenic deposits because they provide, during their deep metamorphic devolatilization, fluids, ligands, and gold. This sediment source model implies that gold is pre-enriched during seafloor sedimentation. It involves the formation of primary nodular pyrite by bacterial activity [65, 99], and gold is incorporated in growing pyrites according to its concentration in the sea waters. Tomkins [92] demonstrated that the solubility of gold in seawater is influenced by the oxidative conditions of the oceans. Based on the composition of the nodular pyrites analyzed by LA-ICP-MS, Large et al. [100] demonstrated that the metal signatures of pyrites in the black shales and their gold concentration varied over geological time, depending on the oxidizing conditions of the oceans [65]. The secular distribution of orogenic deposits across geological times [32, 33] corresponds very well to the periods when the primary pyrites are the richest in gold [65, 100, 101]. Specifically, in greenstone belts between 1800 Ma and 800 Ma, where gold deposits are lacking, the nodular pyrites are poorer in gold. During this period, referred to as the Boring Billion, the oceans are considered to be anoxic and sulfurous [102] and therefore not very favorable for the early gold enrichment of nodular pyrites. These data on nodular pyrites add support to the concept of a sedimentary source for gold [65].

The sediment source model has been independently validated by the presence of ethane in the fluid inclusions from orogenic gold deposits [92]. Ethane can only be produced by the metamorphism of sediments rich in organic matter [23]. The presence of ethane in fluid inclusions has also been documented at the deposit- and district-scale (**Figures 4** and **5**) in the Paleoproterozoic terrain in Burkina Faso [23], Neoproterozoic in Sudan [40], and Mesozoic Otago Schist in New Zealand [41]. For all these terrains, the host sequences are predominantly sedimentary with the presence of pyritic black shales. In short, ethane in fluid inclusions of gold-bearing veins is a proxy for a sedimentary source of gold at depth.

6. Implications for gold exploration

Gold exploration is still based on the adage "Gold is where you find it," as it is driven by gold analyses. The volatile fluid inclusion composition has the potential to be used for gold exploration in orogenic contexts. Previous studies demonstrated the potential of specific volatile composition for vectoring exploration for orogenic gold in metasedimentary sequences [17, 103, 104]. From the various studied cases, the following best signatures of mineralizing fluids can be established.

All samples have CO_2 content, and some are reaching 100 mole% (**Figures 4** and **5**). This is consistent with previous fluid inclusion studies [66, 105], whereas the association of CO_2 with orogenic gold mineralizations is well known. The CO_2 in solution, as carbonic acid, controls the pH in the range of the highest solubility for gold under H_2S-ligands [98, 106]. Consequently, the CO_2 content is critical for gold mineralization and hence provides a key parameter for exploration.

Ethane was established as a proxy for a source of gold from the thermal maturation of organic and pyritic-rich sedimentary rocks at depth. Ethane is thus a key parameter for such an important possible source of gold. In addition, organic matter-rich shales also provide a very important reservoir for carbon dissolved as CO_2 [95] and carbonic acid in solution, as well as light hydrocarbons (CH_4, C_2H_6) and a source for N_2.

Numerous analyzed samples contain H_2S. Sulfur occurs mainly as reduced H_2S species in hydrothermal solutions. Dissolved sulfur is highly important because gold is transported as sulfuric complexes in metamorphic-related gold deposits [96, 97]. H_2S is the main ligand forming Au[HS] complexes [97, 98], and the solubility of Au[HS] is strongly controlled by pH and the oxygen fugacity (ʃO2). Sulfur content of fluid inclusions, especially as reduced species, is thus a premium criterion for the favorability of the hydrothermal solution to carry gold.

Helium is a noble nonreactive element. It is known that He was mainly sourced from mantle outgassing and that transcrustal faults may provide leaking conduits [107]. He was quantified for the HGZ at the Perron deposit and the Detour Gold deposit (**Figure 4A,B**). It is thus suggested that He content of fluid inclusions may be used as an indicator for the proximity of a major fault, a key parameter for the formation of gold deposits in metamorphic terrains [108–110].

Hydrogen is a rare component of the fluid inclusions. It can be involved in various geological reactions, as demonstrated by the gold mineralization in the Otago Schist. Its presence should be considered as indicative of hydrothermal reactions with CO_2, CH_4 and C_2H_6, and hence suggestive of a sedimentary source of gold, as interpreted for the Otago Schist.

Water-poor but organic-rich fluids have the potential of carrying gold under species different from the classical HS-Au complexes. For the HGZ, it is interpreted that gold was transported as hydrocarbon-metal complexes or as colloidal gold nanoparticles [35, 41] and in part as Bi-Te-Au melts [69]. The exceptionally high content of gold is thus attributed to the higher transporting capacity of these unique mineralizing fluids.

7. Future development of the technique

Although volatile contents can yield critical information for gold targeting at various scales, at present, technical issues limit its practical application. The time required for analyzing a sample is the major limiting factor. With the present solid-probe mass spectrometer, only one sample can be analyzed in duplicate and triplicate during the day. Development is ongoing with Laurentia Exploration in Canada to build an apparatus with the capacity to analyze up to 50 samples a day. Commercial services of volatile analysis in fluid inclusions should be soon available.

8. Conclusion

Although there are more modern techniques to analyze fluid inclusions, the solid-mass spectrometric technique offers numerous advantages. The real-time recording of decrepitation ensures that analytical results are not contaminated by synthetic dust, for example, or other contaminants. As decrepitations are directly under vacuum, there is no memory effect or contamination related to gas transfer to the mass spectrometer.

The technique contributed to refining the orogenic gold model. Ethane in fluid inclusions was established as a proxy for a sedimentary source of the gold at depth. This source is now considered one of the most important for forming gold deposits worldwide. Finally, hydrothermal reactions at the district-scale for gold-bearing fluids were documented, including the explanation of the particular water-poor and CO_2-rich fluids commonly associated with world-class gold deposits.

Acknowledgements

The results presented here involved 6 years of apparatus development and 15 years of data collection from gold deposits worldwide. I am indebted to numerous gold exploration companies that provided access to their mines and projects. NSERC and Mitacs provided some financing for specific projects that used the system. Laurentia Exploration is thanked for financing the development of the new analytical system.

DOI: http://dx.doi.org/10.5772/intechopen.1007318

Author details

Damien Gaboury[1,2]

1 University of Québec at Chicoutimi (UQAC), Chicoutimi, Canada

2 Laurentia Exploration, Jonquière, Canada

*Address all correspondence to: dgaboury@uqac.ca;
damien.g@laurentiaexploration.com

References

[1] Chi G, Diamond LW, Lu H, Lai J, Chu H. Common problems and pitfalls in fluid inclusion study: A review and discussion. Minerals. 2020;**11**(1):7

[2] Blamey NJ. Composition and evolution of crustal, geothermal and hydrothermal fluids interpreted using quantitative fluid inclusion gas analysis. Journal of Geochemical Exploration. 2012;**116**:17-27

[3] Chi G, Chou I-M, Lu H-Z. An overview on current fluid-inclusion research and applications. Acta Petrologica Sinica. 2003;**19**:201-212

[4] Roedder E. Fluid inclusion analysis — Prologue and epilogue. Geochimica et Cosmochimica Acta. 1990;**54**:495-507

[5] Caumon M-C, Tarantola A, Wang W. Raman spectra of gas mixtures in fluid inclusions: Effect of quartz birefringence on composition measurement. Journal of Raman Spectroscopy. 2019;**51**:1868-1873

[6] Wagner T, Fusswinkel T, Wälle M, Heinrich CA. Microanalysis of fluid inclusions in crustal hydrothermal systems using laser ablation methods. Elements. 2016;**12**:323-328

[7] Barker C, Smith MP. Mass spectrometric determination of gages in individual fluid inclusions in natural minerals. Analytical Chemistry. 1986;**58**:1330-1333

[8] Bray CJ, Spooner ETC, Thomas AV. Fluid inclusion volatile analysis by heated crushing, on-line gas chromatography; applications to Archean fluids. Journal of Geochemical Exploration. 1991;**42**:167-193

[9] Gleeson SA. Bulk analysis of electrolytes in fluid inclusions. In: Samson I, Anderson A, Marshall D, editors. Fluid Inclusions: Analysis and Interpretation. Vol. 32. Québec, QC, Canada: Mineralogical Association of Canada Publications; 2003. pp. 233-246

[10] Graney JR, Kesler SE. Factors affecting gas analysis of inclusion fluid by quadrupole mass spectrometry. Geochimica et Cosmochimica Acta. 1995;**59**:3977-3986

[11] Kesler SE, Haynes PS, Creech MZ, Gorman JA. Application of fluid inclusions and rock-gas analysis in mineral exploration. Journal of Geochemical Exploration. 1986;**25**:201-215

[12] Landis GP, Rye RO. Characterization of gas chemistry and noble-gas isotope ratios of inclusion fluids in magmatic-hydrothermal and magmatic-steam alunite. Chemical Geology. 2005;**215**:155-184

[13] Norman DI, Sawkins FJ. Analysis of volatiles in fluid inclusions by mass spectrometry. Chemical Geology. 1987;**61**:1-10

[14] Takenouchi S. Fluid inclusion gas composition of some mineral deposits and a geothermal area. Journal of Geochemical Exploration. 1991;**42**:201-215

[15] Guha J, Lu H, Gagnon M. Gas composition of fluid inclusions using solid probe mass spectrometry and its application to study of mineralizing processes. Geochimica et Cosmochimica Acta. 1990;**54**:553-558

[16] Gaboury D, Keita M, Guha J, Lu H-Z. Mass spectrometric analysis of volatiles

in fluid inclusions decrepitated by controlled heating under vacuum. Economic Geology. 2008;**103**:439-443

[17] Chryssoulis SL, Rankin AH. Decrepitometry of fluid inclusions in quartz from the Guadalcazar granite of Mexico; principles and application to mineral exploration. Mineralium Deposita. 1988;**23**:42-49

[18] Leroy J. Contribution à l'échantillonnage de la pression interne des inclusions fluides lors de leur décrépitation. Bulletin de Minéralogie. 1979;**102**:584-593

[19] Neumayr P, Hagemann SG, Banks DA, Yardley BWD, Couture J-F, Landis GP, et al. Fluid chemistry and evolution of hydrothermal fluids in an Archaean transcrustal fault zone network: The case of the Cadillac tectonic zone, Abitibi greenstone belt, Canada. Canadian Journal of Earth Sciences. 2007;**44**:745-773

[20] Norman DI, Musgrave JA. N_2-Ar-He compositions in fluid inclusions: Indicators of fluid source. Geochimica et Cosmochimica Acta. 1994;**58**:1119-1131

[21] Miller PE, Bonner Denton M. The quadrupole mass filter: Basic operating concepts. Journal of Chemical Education. 1986;**64**:617-622

[22] Sauter AD, Downs JJ, Buchner JD, Ringo NT, Shaw DL, Dulak JG. Model for the estimation of electron impact gas chromatography/mass spectrometry response factors for quadrupole mass spectrometers. Analytical Chemistry. 1986;**58**:1665-1670

[23] Gaboury D. Does gold in orogenic deposits come from pyrite in deeply buried carbon-rich sediments?: Insight from volatiles in fluid inclusions. Geology. 2013;**41**:1207-1210

[24] Boullier A-M, Robert F. Palaeoseismic events recorded in Archaean gold-quartz vein networks, Val d'Or, Abitibi, Quebec, Canada. Journal of Structural Geology. 1992;**14**:161-179

[25] Robert F, Kelly WC. Ore-forming fluids in Archean gold-bearing quartz veins at the sigma mine, Abitibi greenstone belt, Quebec, Canada. Economic Geology. 1987;**82**:1464-1482

[26] Robert F, Boullier A-M, Firdaous K. Gold–quartz veins in metamorphic terranes and their bearing on the role of fluids in faulting. Journal of Geophysical Research, Solid Earth. 1995;**100**:12,861-12,879

[27] Kerrich R. Some effects of tectonic recrystallization on fluid inclusions in vein quartz. Contributions to Mineralogy and Petrology. 1976;**59**:195-202

[28] Tarantola A, Diamond LW, Stünitz H, Thust A, Pec M. Modification of fluid inclusions in quartz by deviatoric stress. III: Influence of principal stresses on inclusion density and orientation. Contributions to Mineralogy and Petrology. 2012;**164**:537-550

[29] Goldfarb RJ, Groves DI. Orogenic gold: Common vs evolving fluid and metal sources through time. Lithos. 2015;**223**:2-26

[30] Wilkinson JJ. Fluid inclusions in hydrothermal ore deposits. Lithos. 2001;**55**:229-272

[31] Bradley DC. Secular trends in the geologic record and the supercontinent cycle. Earth Science Reviews. 2011;**108**:16-33

[32] Goldfarb RJ, Bradley D, Leach DL. Secular variations in economic geology. Economic Geology. 2010;**105**:459-466

[33] Tomkins AG. A biogeochemical influence on the secular distribution of orogenic gold. Economic Geology. 2013a;**108**:193-197

[34] Castonguay S, Dubé B, Wodicka N, Mercier-Langevin P. Geological setting and gold mineralization associated with the Sunday Lake and lower detour deformation zones, northwestern Abitibi greenstone belt, Ontario and Quebec. Targeted Geoscience Initiative. 2020;**5**:127-142

[35] Gaboury D, Genna D, Trottier J, Bouchard M, Augustin J, Malcolm K. The Perron gold deposit, Archean Abitibi belt, Canada: Exceptionally high-grade mineralization related to higher gold-carrying capacity of hydrocarbon-rich fluids. Minerals. 2021;**11**(10):1066

[36] Augustin J, Gaboury D. Multi-stage and multi-sourced fluid and gold in the formation of orogenic gold deposits in the world-class Mana district of Burkina Faso - revealed by LA-ICP-MS analysis of pyrites and arsenopyrites. Ore Geology Reviews. 2019;**104**:495-521

[37] Abdelsalam MG, Stern RJ. Mapping Precambrian structures in the Sahara Desert with SIR641 C/X-SAR radar: The Neoproterozoic Keraf suture. NE Sudan: Journal of Geophysical Research: Planets. 1996;**101**:23063-23076. DOI: 10.1029/96JE01391 643

[38] Abdelsalam MG, Stern RJ, Copeland P, Elfaki EM, Elhur B, Ibrahim FM. The Neoproterozoic Keraf suture in Ne Sudan: Sinistral Transpression along the eastern margin of West Gondwana. The Journal of Geology. 1998;**106**:133-148. DOI: 10.1086/516012 648

[39] Bailo T, Schandelmeier H, Franz G, Sun CH, Stern RJ. Plutonic and metamorphic rocks from the Keraf suture (NE Sudan): A glimpse of Neoproterozoic tectonic evolution on the NE margin of W. Gondwana. Precambrian Research. 2003;**123**:67-80. DOI: 10.1016/S0301-9268(03)00044-5

[40] Gaboury D, Nabil H, Ennaciri A, Maacha L. Structural setting and fluid composition of gold mineralization along the central segment of the Keraf suture, Neoproterozoic Nubian shield, Sudan: Implications for the source of gold. International Geology Review. 2022;**64**:45-71. DOI: 10.1080/00206814.2020.1838336

[41] Gaboury D, Mackezie D, Craw D. Fluid volatile composition associated with orogenic gold mineralization, Otago schist, New Zealand: Implications of H_2 and C_2H_6 for fluid evolution and gold source. Ore Geology Reviews. 2021b;**133**:104086. DOI: 10.3390/min11101066

[42] Bierlein FP, Craw D. Petrogenetic character and provenance of metabasalts in the aspiring and Torlesse terranes, South Island, New Zealand: Implications for the gold endowment of the Otago schist? Chemical Geology. 2009;**260**:330344

[43] Mortimer N. Jurassic tectonic history of the Otago schist, New Zealand. Tectonics. 1993;**12**:237244

[44] Mortimer N. Metamorphic discontinuities in orogenic belts: Example of garnet-biotite-albite zone in the Otago schist, New Zealand. International Journal of Earth Sciences. 2000;**89**:295-306

[45] Pitcairn IK, Teagle DAH, Craw D, Olivo GR, Kerrich RT, Brewer TS. Sources of metals and fluids in orogenic gold deposits: Insights from the Otago and alpine schists, New Zealand. Economic Geology. 2006;**101**:1525-1546

[46] Craw D. Geochemistry of late metamorphic hydrothermal alteration and graphitisation of host rock, Macraes gold mine, Otago schist, New Zealand. Chemical Geology. 2002;**191**:257-275

[47] Paterson CJ. Controls on gold and tungsten mineralization in metamorphic-hydrothermal systems, Otago, New Zealand. Geological Association of Canada Special Paper. 1986;**32**:25-39

[48] Williams GJ. Economic Geology of New Zealand. 2nd ed. Vol. 4. The Australasian Institute of Mining and Metallurgy, Monograph. 1974. p. 490

[49] Craw D, MacKenzie D. Macraes Orogenic Gold Deposit (New Zealand): Origin and Development of a World Class Gold Mine. Springer Briefs in World Mineral Deposit. Berlin, Germany: Springer; 2016. p. 123. ISBN 978-3-319-35158-2

[50] Teagle DAH, Norris RJ, Craw D. Structural controls on gold-bearing quartz mineralisation in a duplex thrust system, Hyde Macraes shear zone, Otago schist, New Zealand. Economic Geology. 1990;**85**:1711-1719

[51] MacKenzie DJ, Craw D. Contrasting hydrothermal alteration mineralogy and geochemistry in the auriferous rise & Shine shear zone, Otago, New Zealand. New Zealand Journal of Geology and Geophysics. 2007;**50**(2):67-79

[52] Klein EL, Fuzikawa K. Origin of the CO_2-only fluid inclusions in the Palaeoproterozoic Carará vein-quartz gold deposit, Ipitinga Auriferous District, SE-Guiana shield, Brazil: Implications for orogenic gold mineralisation. Ore Geology Reviews. 2010;**37**:31-40. DOI: 10.1016/j.oregeorev.2009.10.001

[53] Hrstka T, Dubessy J, Zachariáš J. Bicarbonate-rich fluid inclusions and hydrogen diffusion in quartz from the Libčice orogenic gold deposit, bohemian massif. Chemical Geology. 2011;**281**:317-332. DOI: 10.1016/j.chemgeo.2010.12.018

[54] Mironova OF, Naumov VB, Ryzhenko BN. Composition of volatile components in mineral-hosted fluid inclusions at hydrothermal deposits. Water–rock–gas systems in ore-forming processes. Geochemistry International. 2018;**56**:226-233

[55] Kawagucci S et al. Geochemical origin of hydrothermal fluid methane in sediment-associated fields and its relevance to the geographical distribution of whole hydrothermal circulation. Chemical Geology. 2013;**339**:213-225. DOI: 10.1016/j.chemgeo.2012.05.003

[56] Whiticar MJ. Carbon and hydrogen isotope systematics of bacterial formation and oxidation of methane. Chemical Geology. 1999;**161**:291-314. DOI: 10.1016/S0009-2541(99)00092-3

[57] McCollom TM, Seewald JS. Abiotic synthesis of organic compounds in deep sea hydrothermal environments. Chemical Reviews. 2007;**107**:382-401. DOI: 10.1021/cr0503660

[58] Schrenk MO, Brazelton WJ, Lang SQ. Serpentinization, carbon, and deep life. Reviews in Mineralogy and Geochemistry. 2013;**75**:575-606. DOI: 10.2138/rmg.2013.75.18

[59] McCollom TM. Laboratory simulations of abiotic hydrocarbon formation in Earth's deep subsurface. Reviews in Mineralogy and Geochemistry. 2013;**75**:467-494. DOI: 10.2138/rmg.2013.75.15

[60] Chi G, Dubé B, Williamson K, Williams-Jones AE. Formation of the Campbell-red lake gold deposit by H_2O-poor, CO_2-dominated fluids. Mineralium Deposita. 2006;**40**:726-741

[61] Chi G, Liu Y, Dubé B. Relationship between CO_2-dominated fluids, hydrothermal alterations and gold mineralization in the red Lake greenstone belt, Canada. Applied Geochemistry. 2009;**24**:504-516

[62] Klemd R, Hirdes W. Origin of an unusual fluid composition in early Proterozoic Palaeoplacer and lode-gold deposits in Birimian greenstone terranes of West Africa. South African Journal of Geology. 1997;**100**:405-414

[63] Schmidt Mumm A, Oberthür T, Vetter U, Blenkinsop TG. High CO_2 content of fluid inclusions in gold mineralizations in the Ashanti Belt, Ghana: A new category of ore forming fluids? Mineralium Deposita. 1997;**32**:107-118

[64] Schmidt Mumm A, Oberthür T, Vetter U, Blenkinsop TG. High CO_2 content of fluid inclusions in gold mineralizations in the Ashanti Belt, Ghana: A new category of ore forming fluids? – A reply. Mineralium Deposita. 1998;**33**:320-322

[65] Gaboury D. The neglected involvement of organic matter in forming large and rich hydrothermal orogenic gold deposits. Geosciences. 2021;**11**(8):344

[66] Ridley JR, Diamond LW. Fluid chemistry of orogenic lode gold deposits and implications for genetic models. Reviews in Economic Geology. 2000;**13**:146-162

[67] Goldfarb RJ, André-Mayer A-S, Jowitt SM, Mudd GM. West Africa: The World's premier paleoproterozoic gold province. Economic Geology. 2017;**112**:123-143

[68] Lüders V, Klemd R, Oberthür T, Plessen B. Different carbon reservoirs of auriferous fluids in African Archean and Proterozoic gold deposits? Constraints from stable carbon isotopic compositions of quartz-hosted CO_2-rich fluid inclusions. Mineral Deposits. 2015;**50**:449-454

[69] Gaboury D, Genna D, Augustin J, Bouchard M, Trottier J. LA-ICP-MS analyses of sulfides from gold-bearing zones at the Perron deposit, Abitibi Belt, Canada: Implications for gold remobilization through metamorphism from volcanogenic Mineralizations to orogenic quartz–carbonate veins. Minerals. 2024;**14**:843. DOI: 10.3390/min14080843

[70] Crede LS, Evans KA, Rempel KU, Brugger J, Etschmann B, Bourdet J, et al. Revisiting hydrocarbon phase mobilization of Au in the Au–Hg McLaughlin mine, geysers/clear Lake area, California. Ore Geology Reviews. 2020;**117**:103218

[71] Emsbo P, Koenig AE. Transport of Au in petroleum: Evidence from the Northern Carlin trend, Nevada. In: Mineral Exploration and Research: Digging Deeper. Proceedings of the 9th Biennial SGA Meeting. Dublin: Millpress; 2007. pp. 695-698

[72] Bul'bak TA, Tomilenko AA, Gibsher NA, Sazonov AM, Shaparenko EO, Ryabukha MA, et al. Hydrocarbons in fluid inclusions from native gold, pyrite, and quartz of the Sovetskoe deposit (Yenisei ridge, Russia) according to pyrolysis-free gas chromatography-mass spectrometry data. Russian Geology and Geophysics. 2020;**61**(11):1260-1282

[73] Fuchs S, Schumann D, Williams-Jones AE, Vali H. The growth and concentration of uranium and titanium minerals in hydrocarbons of the carbon leader reef, witwatersrand supergroup,

South Africa. Chemical Geology. 2015;**393-394**:55-66

[74] Fuchs S, Williams-Jones AE, Jackson SE, Przybylowicz WJ. Metal distribution in pyrobitumen of the carbon leader reef, witwatersrand supergroup, South Africa: Evidence for liquid hydrocarbon ore fluids. Chemical Geology. 2016;**426**:45-59

[75] Crede LS, Evans KA, Rempel KU, Grice K, Sugiyama I. Gold partitioning between 1-dodecanethiol and brine at elevated temperatures: Implications of Au transport in hydrocarbons for oil-brine ore systems. Chemical Geology. 2019a;**504**:28-37

[76] Crede LS, Liu W, Evans KA, Rempel KU, Testemale D, Brugger J. Crude oils as ore fluids: An experimental in-situ XAS study of gold partitioning between brine and organic fluid from 25 to 250°C. Geochimica et Cosmochimica Acta. 2019b;**244**:352-365

[77] Ding Z, Sun X, Xu X, Ling S, Xie Z. Likely gold transportation via organic complexes in sediment-hosted orogenic gold deposits in southern Tibet: From the perspective of thermodynamic modelling. Journal of Asian Earth Sciences. 2024;**264**:106068

[78] McLeish DF, Williams-Jones AE, Vasyukova OV, Clark JR, Board WS. Colloidal transport and flocculation are the cause of the hyperenrichment of gold in nature. Proceedings of the National Academy of Sciences. 2021;**118**(20):e2100689118

[79] Petrella L, Thébaud N, Fougerouse D, Evans KQ, Z. & Laflamme, C. Colloidal gold transport: A key to high-grade gold mineralization. Mineral Deposits. 2020;**55**:1247-1254

[80] Saunders JA. Colloids and nanoparticles: Implications for hydrothermal precious metal ore formation. SEG Newsletter. 2022;**130**:15-21

[81] Frimmel HE. Earth's continental crustal gold endowment. Earth and Planetary Science Letters. 2008;**267**:45-55

[82] Gaboury D. Parameters for the formation of orogenic gold deposits. Applied Earth Science. 2019;**128**:124-133

[83] Sillitoe RH, Thompson JF. Intrusion–related vein gold deposits: Types, tectono-magmatic settings and difficulties of distinction from orogenic gold deposits. Resource Geology. 1998;**48**(4):237-250

[84] Augustin J, Gaboury D. Paleoproterozoic plume-related basaltic rocks in the Mana gold district in western Burkina Faso, West Africa: Implications for exploration and the source of gold in orogenic deposits. Journal of African Earth Sciences. 2017;**129**:17-30

[85] Patten CGC, Pitcairn IK, Molnár F, Kolb J, Beaudoin G, Guilmette C, et al. Gold mobilization during metamorphic devolatilization of Archean and Paleoproterozoic metavolcanic rocks. Geology. 2020;**48**:1110-1114. DOI: 10.1130/G47658.1

[86] Pitcairn IK, Craw D, Teagle DAH. Metabasalts as sources of metals in orogenic gold deposits. Mineral Deposits. 2015;**50**:373-390

[87] Large RR, Bull SW, Maslennikov VV. A carbonaceous sedimentary source-rock model for Carlin-type and orogenic gold deposits. Economic Geology. 2011;**106**:331-358

[88] Large R, Thomas H, Craw D, Henne A, Henderson S. Diagenetic pyrite as a source for metals in orogenic gold

deposits, Otago schist, New Zealand. New Zealand Journal of Geology and Geophysics. 2012;**55**:137-149

[89] Pitcairn IK, Olivo GR, Teagle DA, Craw D. Sulfide evolution during prograde metamorphism of the Otago and alpine schists, New Zealand. The Canadian Mineralogist. 2010;**48**(5):1267-1295

[90] Pitcairn IK, Leventis N, Beaudoin G, Faure S, Guilmette C, Dubé B. A metasedimentary source of gold in Archean orogenic gold deposits. Geology. 2021;**49**(7):862-866. DOI: 10.1130/G48587.1

[91] Suhendra R, Takahashi R, Agangi A, Imai A, Sato H, Setiawan NI. Textural-compositional evolution of pyrite and metal remobilization during low-grade metamorphism of metapelite: Contribution to gold mineralization in the Luk Ulo complex, Central Java, Indonesia. Ore Geology Reviews. 2014;**166**:105966

[92] Tomkins AG. On the source of orogenic gold. Geology. 2013b;**41**:1255-1256

[93] Thomas HV, Large RR, Bull SW, Maslennikov V, Berry RF, Fraser R, et al. Pyrite and pyrrhotite textures and composition in sediments, laminated quartz veins, and reefs at Bendigo gold mine, Australia: Insights for ore genesis. Economic Geology. 2011;**106**:1-31

[94] Finch EG, Tomkins AG. Pyrite-pyrrhotite stability in a metamorphic aureole: Implications for orogenic gold genesis. Economic Geology. 2017;**112**:661-674

[95] Tomkins AG. Windows of metamorphic sulfur liberation in the crust: Implications for gold deposit genesis. Geochimica et Cosmochimica. 2010;**74**:3246-3259

[96] Stefánsson A, Seward TM. Gold(I) complexing in aqueous sulphide solutions to 500°C at 500 bar. Geochimica et Cosmochimica Acta. 2004;**68**:4121-4143

[97] Williams-Jones AE, Bowell RJ, Migdisov AA. Gold in solution. Elements. 2009;**5**:281-287

[98] Phillips GN, Powell R. Formation of gold deposits: A metamorphic devolatilization model. Journal of Metamorphic Geology. 2010;**28**:689-718

[99] Rickard D, Mussmann M, Steadman JA. Sedimentary sulfides. Elements. 2017;**13**:117-122

[100] Large RR, Gregory DD, Steadman JA, Tomkins AG, Lounejeva E, Danyushevsky LV, et al. Gold in the oceans through time. Earth and Planetary Science Letters. 2015;**428**:139-150

[101] Steadman JA, Large RR, Blamey NJ, Mukherjee I, Corkrey R, Danyushevsky LV, et al. Evidence for elevated and variable atmospheric oxygen in the Precambrian. Precambrian Research. 2020;**343**:105722

[102] Canfield D. A new model for Proterozoic ocean chemistry. Nature. 1998;**396**:450-453

[103] Kerr MJ, Hanley JJ, Kontak DJ. Application of bulk fluid volatile chemistry to exploration for metasedimentary rock-hosted orogenic gold deposits: An example from the Meguma terrane, Nova Scotia, Canada. Journal of Geochemical Exploration. 2021;**226**:106764

[104] Shepherd TJ, Bottrell SH, Miller MF. Fluid inclusion volatiles as an exploration

guide to black shale-hosted gold deposits, Dolgellau gold belt, North Wales, UK. Journal of Geochemical Exploration. 1991;**42**(1):5-24

[105] Prokofiev VY, Naumov VB. Physicochemical parameters and geochemical features of ore-forming fluids for orogenic gold deposits throughout geological time. Minerals. 2020;**10**(1):50. DOI: 10.3390/min10010050

[106] Phillips GN, Evans KA. Role of CO_2 in the formation of gold deposits. Nature. 2004;**429**:860-863

[107] Lupton JE, Weiss RF, Craig H. Mantle helium in hydrothermal plumes in the Galapagos rift. Nature. 1977;**267**:603-604

[108] Bateman R, Bierlein FP. On Kalgoorlie (Australia), Timmins porcupine (Canada), and factors in intense gold mineralisation. Ore Geology Reviews. 2007;**32**:187-206

[109] Bedeaux P, Pilote P, Daigneault R, Rafini S. Synthesis of the structural evolution and associated gold mineralization of the Cadillac fault, Abitibi, Canada. Ore Geology Reviews. 2017;**82**:49-69

[110] Dubé B, Gosselin P. Mineral deposits of Canada: A synthesis of major deposit-types, district metallogeny, the evolution of geological provinces, and exploration methods. Geological Association of Canada, Mineral Deposits Division, Special Publication. 2007;**5**:49-73

Chapter 4

Solid Solution Thermochemistry: Understanding Metal Zoning in Polymetallic Hydrothermal Sulfide Ore Deposits

Richard Sack

Abstract

This chapter is a tutorial illustrating the methods used to create thermochemical models for sulfide solid solutions focusing on the sulfosalt fahlore ≈ $(Cu,Ag)_{10}(Fe,Zn)_2(Sb,As)_4S_{(13-x)}$ with S vacancies $0 \leq x < 3$ and the common sulfide sphalerite (Zn,Fe)S. The formulation and calibration of the thermochemical model for sulfur vacancy-free fahlores are outlined and preliminary estimates for the Gibbs energies of formation of As-endmembers are presented. For sphalerites it summarizes the development of a cluster variation method (CVM) model based on 13 and 6 metal atom cubeocahedral and octahedral basis clusters which accounts for experimental constraints on sphalerite compositions in pyrrhotite + pyrite and pyrrhotite + iron metal assemblages, calorimetric and 850°C activity-composition constraints, and *ab-initio* quantum mechanical calculations. The analysis outlines the need for additional theoretical, experimental and petrological studies to complete the model for fahlores with S vacancies, discusses why there is an intrinsic limit of 3 sulfur vacancies in fahlores, and reviews how the large Gibbs energies of the reciprocal reactions governing the incompatibilities between Zn and As, Zn and Ag, and As and Ag are the keys to understanding the origin of metal zoning of fahlores in fissure-vein, polymetallic sulfide ore deposits.

Keywords: thermochemistry, fahlore, sphalerite, sulfur vacancies, metal zoning

1. Introduction

Metal zoning in polymetallic sulfide ore deposits of fissure-vein type is so common that [1] noted that its absence is an exception to the rule. Historically such zoning has been documented with metal ratios obtained from assay data, such as the Pb/Cu and Ag/Pb ratios employed by Refs. [2, 3] to illustrate prominent lateral variations in the chemistry of such veins over more than 4 km in the Herminia Mine from the south eastern portion of the Julcani mining district (Angaraes, Peru), (compare **Figure 1**). In this mine at the 330 level in 1975, the Pb/Cu ratio increased progressively along the direction of fluid flow, rising from less than 1 in the northwest to >50 over 4 km

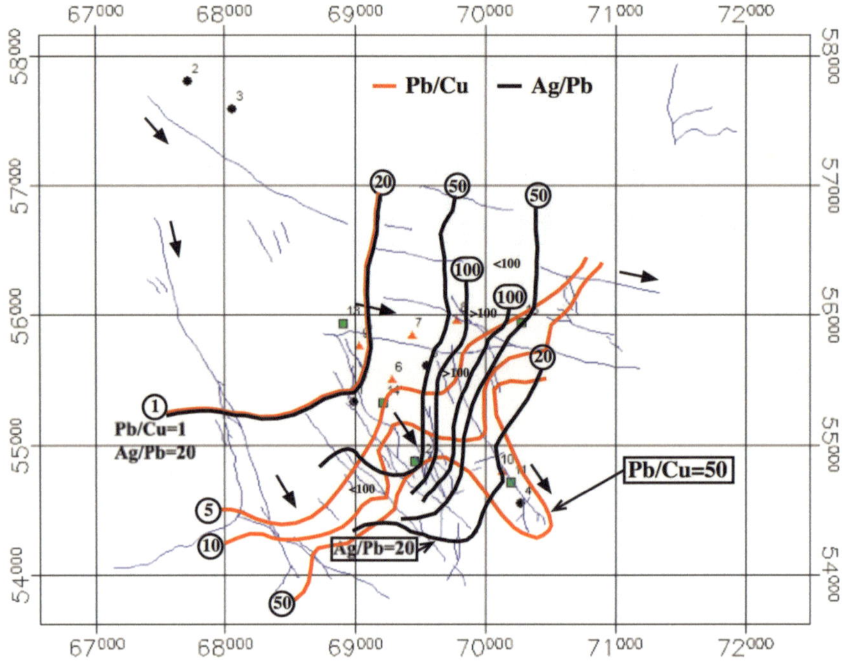

Figure 1.
Contours of Pb/Cu and Ag (ounces per ton)/Pb (wt %) metal ratios of metal veins (blue lines) in the Herminia mine as they appeared at the 330 level in 1975. Arrows indicate inferred directions of hydrothermal fluid flow; symbols with numbers indicate locations of samples examined by Sack and Goodell [4]. From Sack and Goodell [4].

downstream to the southeast, but the Ag (ounces/ton)/Pb (wt %) ratio underwent a maximum above 100 at the apex of a zone of massive Ag dumping, and then came back down further downstream to the southeast. It is in the downstream halo of the zone of Ag dumping that samples with textural evidence for primary galenas containing more than 80 mole % Ag(Sb,Bi)S_2 were found [4]. Although this type of analysis may be useful for mineral exploration and mine development, it does not necessarily aid in the understanding of paragenesis, because such ratios are often the product of several superimposed mineralizations.

As an alternative, variations in the compositions of solid solutions such as sphalerite and fahlore may be used to outline the evolution of physicochemical parameters of the hydrothermal fluids with respect to time and space. Of course, such an approach requires that the primary depositional compositions of the minerals are preserved or may be reconstructed, and coeval spatial horizons in the veins may be identified. Given that sphalerite is quite refractory and its color is very sensitive to FeS content [5], it is not surprising that studies of sphalerite compositions in hydrothermal veins were first conducted for this purpose [6]. And more recently there have been attempts to use fahlore for this purpose (for example, see [7–10]).

Because of the well preserved stratigraphy of sulfides in the metal veins, Refs. [7, 8] were able to demonstrate that fahlores from the Casapalca Pb-Zn-Cu-Ag mine (near the crest of the Peruvian Andes) were zoned in As/(As+ Sb) and Ag/(Ag + Cu) in both time and space. At Casapalca fahlore occurs in veins contained in a stockwork-like fracture system [11] in a well defined stratigraphic sequence in which fahlore is preceded by more volumetrically abundant sphalerite, galena and pyrite, follows closely associated chalcopyrite, and is in turn followed by bournonite and the gangue minerals calcite and dolomite. Quartz crystallizes throughout all sulfide precipitating stages of mineralization, and realgar, orpiment, stibnite and other Pb-Ag sulfosalts occur in the upper levels of the mine whose metal veins extend over 5 km laterally, extend over 2 km vertically down their near vertical dip, and are continuous, persistent and connected. Alteration of the wall rocks (primarily Tertiary red beds of shale, sandstone and some limestone, andesitic tuffs, porphyries, breccias and conglomerates) extends several hundred meters to less than a few centimeters from the veins from the core to the distal portions of the deposit, and results in finely-grained silica-rich rocks which are sericitizied, pyritizied and propylitizied. Crystallization temperatures established from fluid inclusions indicate that fahlores crystallized at temperatures between about 300 and 280°C, while the early formed sulfides and post sulfide carbonates precipitated at temperature of around 370 and 200°C, respectively [12]. Finally, H and O isotopes indicate that ore deposition may be explained by a simple model with meteoric water input virtually nonexistent until carbonate mineralization, relatively uniform S and C isotopic compositions indicate that the pH and f_{O_2} of the ore fluid are consistent with the inference that H_2S comprised more than 90% of the S in the hydrothermal fluid, and fluid inclusions indicate that no significant temperature gradients existed throughout the mine at a given period of mineralization [8, 12, 13].

Wu and Petersen [8] document that the first precipitated fahlores are the most Sb-rich and that these increase in both Sb/(Sb + As) and Ag/(Ag + Cu) both vertically and laterally away from the core of the deposit (compare **Figure 2**). Subsequent fahlores become more As-rich and Ag-poor at a given horizon, and correlation bands from different horizons define the envelopes of early and late fahlore compositions over roughly 1500 m vertically. Trends in Zn/(Zn + Fe) in fahlore are less apparent with this ratio approximating 0.75 for most samples, but more Fe-rich fahlores are produced at least locally, possibly as a result of replacement of chalcopyrite by pyrite in a complex reaction such as that suggested by Wu and Petersen [8]:

$$\begin{aligned} &10CuFeS_2 + xFe^{2+} + (2-x)Zn^{2+} + yAs^{3+} + (4-y)Sb^{3+} + 13H_2S \\ &\Rightarrow 10FeS_2 + Cu_{10}(Fe,Zn)_2(Sb,As)_4S_{13} + 5H_2 + 16H^+ \end{aligned} \tag{1}$$

where $0 \le x \le 2$ and $0 \le y \le 4$, and [8] cited ample evidence for coupling between Sb/(Sb + As) and Ag/(Ag + Cu) ratios of fahlores in other polymetallic sulfide ore deposits.

Hackbarth and Petersen [9] attempted to draw similar envelopes for fahlores for the Ag-Pb-Zn ores from the Coeur d'Alene district (Idaho, USA). Unfortunately, the stratigraphy of sulfide deposition is not typically preserved in the extensively deformed metal veins from the Coeur d'Alene mining district, so they attempted to draw them to be consistent with the assumption that the Ag-rich fahlores observed in the district were the product of metal fractionation between the fahlore and hydrothermal fluid. Unfortunately the empirical fractionation model they developed to rationalize their inferred envelopes did not account for the very nonideal thermodynamic mixing properties of fahlore. Sack [14] attempted to correct for this deficiency in their model by taking into account these properties, and noted that much of their inferred fahlore composition envelopes could be explained by the non ideal thermodynamic mixing properties of fahlore. However, subsequent studies (for example, [15, 16]) demonstrated that the Ag-enrichment observed in Coeur d'Alene fahlores is a consequence of the retrograde reaction

$$\underset{\text{Galena}}{AgSbS_2} + \underset{\text{Galena}}{PbS} \rightleftharpoons \underset{\text{Fahlore}}{Ag(Cu)_{-1}} + \underset{\text{Bournonite}}{CuPbSbS_3}, \tag{2}$$

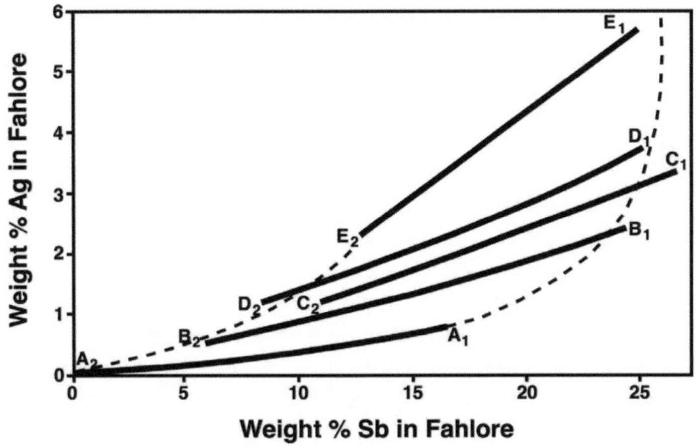

Figure 2.
Antimony-silver correlations for a number of samples represented as line segments. Each line segment (A–E) is a composite of several samples and the subscripts 1 and 2 indicate the first and last deposited fahlores for these line segments. From Wu [7].

whereby the primary $AgSbS_2$ content of fickle galena is removed during cooling, producing the population of high Ag fahlores by Ag for Cu exchange, the bournonite stage of mineralization, and the virtualy $AgSbS_2$-free galenas today present in the district (for example, [17]) and in nearly all galenas in sulfide ore deposits near the surface of the planet Earth. And similar issues are also associated with chalcopyrite, another mineral solid solution whose primary compositons are seldom preserved (for example, [18, 19]).

The Ag-rich Sb-fahlores in the Ag-Pb-Zn ores from the Keno Hill mining district (Yukon, CA; [10, 18, 20–22]) provide another compelling example of prominent primary metal zoning over a scale of 15–20 km from east to west, with six out of eight mines with fahlores with the highest average Ag/(Ag + Cu) ratios with decreasing Zn/(Zn + Fe) ratios, the Porcupine, Shamrock, Lucky Queen, Ruby, Dixie, Husky, HR-224 and Silver King mines, having Ag/(Ag + Cu) and Zn/(Zn + Fe) ratios plotting between the 250 and 310°C isotherms for Sb-fahlores in equilibrium with sphalerite + pyrargyrite [$(Ag,Cu)_3SbS_3$] + miargyrite [$(Ag,Cu)SbS_2$] ([23]; compare **Figure 3**), the temperature range established for primary fahlore mineralization by fluid inclusions in quartz and siderite [27]. This parallelism between average fahlore compositions for these mines and the 250–310°C fahlore isotherms suggests that the primary galenas had substantially higher $AgSbS_2$ contents (and $AgSbS_2$ activities approaching those of miargyrite) than their currently negligible values (compare [15, 16, 18, 21]). This implies that most of the $AgSbS_2$ contents of the primary galenas were subsequently released during cooling of these mesothermal ores and provided most of the elements for epithermal mineralization of stephanite [$(Ag,Cu)_5SbS_4$], polybasite [$(Ag,Cu)_{16}Sb_2S_{11}$], argentite [$(Ag,Cu)_2S$] and wire Ag, minerals of progressively decreasing Sb/Ag ratios relative to miargyrite that are consistent with concentration of Sb in the extreme distal portions of the Keno Hill system during this epithermal mineralization. And such differential transport of semimetals to distal portions of the mineralization is also commonly observed in other Ag-Pb-Zn ore districts such as the Coeur d'Alene (Idaho, USA), and the Casapalca and Julcani districts (Angaraes, Peru) (for example, [2, 4]).

The fahlores from the the Keno Hill mining district are also of interest, because some of them provided the first confirmation of the existence of miscibility gaps in

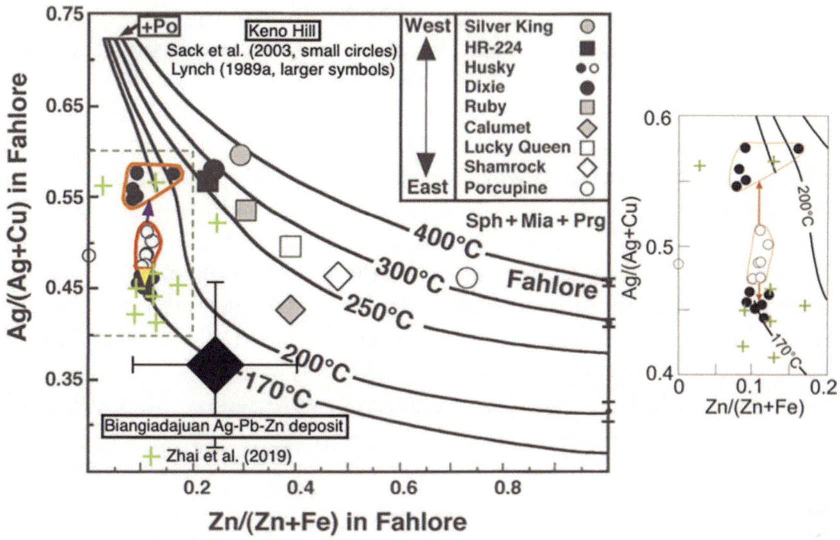

Figure 3.
Isotherms for the molar Ag/(Ag + Cu) and Zn/(Zn + Fe) ratios of $(Cu,Ag)_{10}(Fe,Zn)_2Sb_4S_{13}$ fahlores in the miargyrite + pyrargyrite + sphalerite assemblage [23] compared with mine average fahlore data [10] and the global average for fahlores from mines with ores displaying significant retrograde re-equilibration (large black diamond) and determinations for homogeneous and unmixed fahlores (small white and black circles) from the Keno Hill mining district (Yukon, Canada) of Refs. [10, 21]. The data for primary fahlores from the Biangiadajuan Ag-Pb-Zn deposit (Inner Mongolia, China) ([24], green crosses) are shown for comparison. Arrows attached to the lower red polygon indicate the sense of unmixing of homogeneous portions of the grain examined by Sack et al. [21] into relatively Ag-rich and Ag-poor compositions, points further clarified by the expansion of the composition field for fahlores with Ag/(Ag + Cu) and Zn/(Zn + Fe) ratios between 0.4 and 0.6 and 0.0 and 0.2 that appears to the right of the main figure. Dots and 1σ error bars on the vertical axis for Zn/(Zn + Fe) = 1.00 represent means and standard deviations of molar Ag/(Ag + Cu) ratios of $(Cu,Ag)_{10}Zn_2Sb_4S_{13}$ fahlores at 200, 300, and 400°C determined in the experiments of Ebel and Sack [25]. From Sack et al. [26].

$(Cu,Ag)_{10}(Fe,Zn)_2Sb_4S_{13}$ fahlore first predicted as a result of the thermodynamic analysis by O'Leary and Sack [28] of the Fe-Zn partitioning between fahlores approximating these compositions and sphalerites approximating (Zn,Fe)S in composition. Sack et al. [21] found a grain of fahlore from the Husky mine exhibiting evidence of unmixing of homogeneous regions into Ag-rich and Ag-poor compositions with molar Ag/(Ag + Cu) and Zn/(Zn + Fe) ratios near those of the binodes of the 170°C fahlore isotherm in **Figure 3**. This approximate 170°C temperature for the unmixing is roughly consistent with the temperature inferred by averaging the molar Ag/(Ag + Cu) and Zn/(Zn + Fe) ratios obtained by Lynch [10] for fahlores from the mines exhibiting the greatest extent of retrograde re-equilibration, the Homestake, Onek, Sadie and Husky mines, and comparing these averages, 0.367 ± 0.090 and 0.244 ± 0.159 with the calculated fahlore isotherms in **Figure 3**. It is also consistent with the presence of stephanite (T < 197°C, [29]) in the sulfide assemblage coexisting with the grain of fahlore from the Husky mine exhibiting this evidence of unmixing that was examined by Sack et al. [21]. Despite this confirmation of the prediction of these miscibility gaps based on a thermodynamic analysis of Fe-Zn partitioning between fahlore and sphalerite, this confirmation has been virtually ignored by the mineralogical community, probably because it conflicts with their preconceived notions about the structural role of Ag in these fahlores (for example, [30], p. 54). Although this continues to be the case, the physical evidence for these gaps continues to mount (for example, [24, 31–33]), with the composition data determined by Zhai et al. [24] for the primary fahlores from the Biangiadajuan Ag-Pb-Zn deposit (Inner Mongolia, China) being especially clear and unequivocal. As may be readily inferred from **Figure 3**, these fahlores effectively map out the miscibility gap relations inferred by Refs. [19, 23] for temperatures inferred from the fahlore isotherms and those inferred from previous fluid inclusion studies [34, 35] which indicate that the temperatures for Ag-dominant sulfosalt formation were between ≈160–250°C and <150°C for final calcite formation.

In the case of Keno Hill, the mechanism of ore formation is well known. Oxidized hydrothermal fluids emanating from the plutonic source to the east (the Mayo Lake pluton) encountered a graphitic quartzite (the Keno Hill quartzite), reaction between the graphite and the hydrothermal fluid converted SO_4^{2-} to S^{2-} in the fluid leading to the precipitation of the metal sulfides, and the zoning pattern reflects the incompatibly between Ag and Zn in the fahlore structure as expressed by the positive Gibbs energy of the reciprocal reaction

$$Cu_{10}Zn_2Sb_4S_{13} + Ag_{10}Fe_2Sb_4S_{13} = Cu_{10}Fe_2Sb_4S_{13} + Ag_{10}Zn_2Sb_4S_{13}, \quad (3)$$

$\Delta\overline{G}^o_{24}$, with a positive value of $\ln[(n_{Ag}/n_{Cu})^{FLUID}(n_{Cu}/n_{Ag})^{FAHLORE}]$ leading to downstream enrichment of Ag in fahlore (compare [18]). In the case of Casapalca, the mechanism of ore formation is not as transparent, but it seems more than likely that the spatial composition paths of fahlore may be explained by the strong incompatibly between Ag and As as expressed by the large Gibbs energy of reciprocal reaction:

$$Cu_{10}Fe_2As_4S_{13} + Ag_{10}Fe_2Sb_4S_{13} = Cu_{10}Fe_2Sb_4S_{13} + Ag_{10}Fe_2As_4S_{13}, \quad (4)$$

$\Delta\overline{G}^o_{34}$. It is these matters that we will examine further utilizing recent advances in solid solution theory. These advances will help enable future calibrations of solid–aqueous solution equilibria paralleling those of Refs. [18, 22] and integrating such calibrations into models of reactive transport (for example, [36]). Before reviewing these advances in solid solution theory it is appropriate that we first review the equations used to derive the chemical potentials of endmember components in complex solid solutions.

2. Solid solution thermochemistry

2.1 Binary solutions

The molar Gibbs energy of mixing of a binary solution may be expressed as a function of one independent mole fraction. Where there is a single substitution such as that in an (A,B) or (A,B)X solid solution between endmembers with the same structures, the substitution is limited to only one type of crystallographic site, and there is a random distribution of A and B on this site in the crystalline lattice, the thermodynamic mixing properties of a crystal may typically be described quite accurately as a simple solution with this mole fraction using the formalism developed by Margules [37] and later modified by Refs. [38–42] and others. And the parameters of resulting expressions (for example, symmetric or asymmetric regular solution parameters) may exhibit very minor or negligible dependence on temperature and pressure, particularly when the size differences between A and B are small to negligible (for example, [43]). But for binary solutions involving coupled substitutions (for example, plagioclase feldspar) or exhibiting long- or short-range ordering, more complex treatments are required and these may lead to much more complicated expressions, and in some cases unrealistic temperature dependent parameters, when their solution properties are expressed in terms of this independent mole fraction. We will restrict our attention to these later cases, as simple solutions are more than adequately covered in the papers mentioned above and many other papers and books.

To simplify the discussion we will consider binary solutions of increasing complexity resulting from long-range ordering of cations between crystallographically distinct sites and from short-range ordering. The simplest case of long-range ordering we will consider is $(Mg, Fe)_2Si_2O_6$ orthopyroxene where Fe and Mg fully occupy two sites of equal multiplicity, typically designated as M2 and M1 (for example, [44–47]). We next consider fahlores on the join $Cu_{10}Fe_2Sb_4S_{13} - Ag_{10}Fe_2Sb_4S_{13}$. In these fahlores Ag and Cu are ordered between trigonal-planar and tetrahedral metal sites, but the sums of Ag and Cu are different on these sites, with Ag and Cu fully occupying the 6 trigonal-planar metal sites, but only occupying $\frac{2}{3}$ of the six tetrahedral metal sites they share with two Fe in the structural formula $(Cu, Ag)_6^{TRG}\left([Cu, Ag]_{2/3}Fe_{1/3}\right)_6^{TET} Sb_4S_{13}$ (for example, [19]). Finally, we will consider (Zn,Fe)S sphalerites, where it is necessary to consider short-range ordering of Fe and Zn in various cluster configurations to account for its physical, magnetic and thermodynamic properties (for example, [48, 49]). Before proceeding with this analysis, we will briefly review the equations that are commonly employed in deriving the chemical potentials of the physical endmembers of multicomponent systems with simple solid solutions (i.e. those without any of the complicating factors listed above).

For a simple binary solid solution 1–2 with no long- or short-range ordering of atoms on distinct crystallographic sites the chemical potentials of the endmember components 1 and 2 may be derived from the tangent intercept rule relating the tangent to the molar Gibbs energy surface of the solid solution at a given composition, temperature and pressure to its intercepts on the Gibbs energy axes for the endmember components:

$$\mu_i = \overline{G} + (1 - X_i)\left(\frac{\partial \overline{G}}{\partial X_i}\right)_{T,P}, \tag{5}$$

where $\overline{G}$ is the molar Gibbs energy and the mole fractions of the endmember components 1 and 2, with their mole fractions, X_1 and X_2, summing to unity. Application of this equation thus yields the results

$$\mu_1 = \overline{G} - (X_2)\left(\frac{\partial \overline{G}}{\partial X_2}\right)_{T,P} \tag{6}$$

and

$$\mu_2 = \overline{G} + (1 - X_2)\left(\frac{\partial \overline{G}}{\partial X_2}\right)_{T,P}, \tag{7}$$

when it is recognized that $(1 - X_1) = X_2$. The molar Gibbs energy of the solid solution at a given composition is thus given by the relation

$$\overline{G} = \mu_1(1 - X_2) + \mu_2(X_2). \tag{8}$$

Eq. (5) may be extended to simple higher dimensional multicomponent solutions employing the expression

$$\mu_i = \overline{G} + (1 - X_i)\left(\frac{\partial \overline{G}}{\partial X_i}\right)_{X_j/X_k,T,P} \tag{9}$$

(e.g. Darken [50]), an expression in which the ratios X_j/X_k are held constant during variations of X_i and which thus has the conditions

$$\frac{\partial X_j}{\partial X_i} = \frac{-X_j}{(1 - X_i)} \tag{10}$$

and

$$\frac{\partial X_k}{\partial X_i} = \frac{-X_k}{(1 - X_i)} \tag{11}$$

for a ternary solid solution.

2.1.1 $(Mg, Fe)_2Si_2O_6$ orthopyroxenes

For $(Fe, Mg)^{M2}(Mg, Fe)^{M1}Si_2O_6$ orthopyroxenes we may readily develop expressions for the chemical potentials which make explicit provision for their dependence on the energy of disordering, and on cooperative effects such as any departure of the endmembers $Mg_2Si_2O_6$ and $Fe_2Si_2O_6$, and the ordered and anti-ordered intermediate compositions $(Fe)^{M2}(Mg)^{M1}Si_2O_6$ and $(Mg)^{M2}(Fe)^{M1}Si_2O_6$ from coplanarity in their Gibbs energy (compare **Figure 4**). To do so we may start by treating these pyroxenes

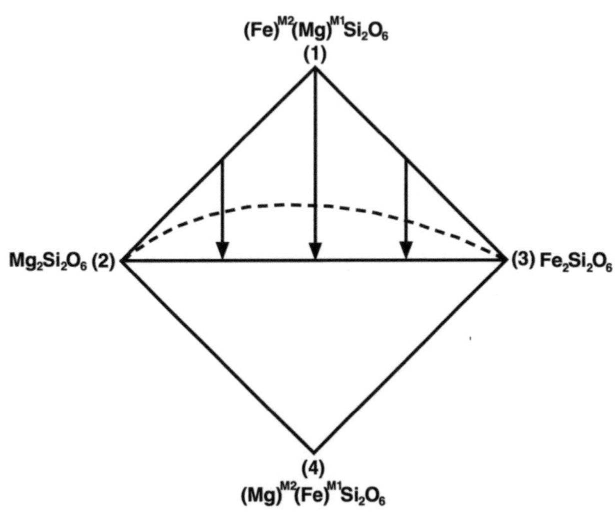

Figure 4.
Composition-ordering space for $(Mg, Fe)_2Si_2O_6$ orthopyroxene. The top half of the diamond defines the accessible portion of this composition-ordering space, with vertical lines with downward pointing arrows indicating variations from complete Fe-Mg ordering to complete disordering on M2 and M1 sites for given bulk Fe/(Fe + Mg) ratios. Dashed curve is a schematic equilibrium ordering-composition path for an arbitrary temperature (compare [47], fig. 20a).

as if they were ternary reciprocal solutions and define the composition space in terms of the mole fractions of the ordered and endmember components:

$$X_1 = (Fe)^{M2}(Mg)^{M1}Si_2O_6, \tag{12}$$

$$X_2 = Mg_2Si_2O_6, \tag{13}$$

$$X_3 = Fe_2Si_2O_6 \tag{14}$$

where

$$X_{Mg}^{M2} = X_2 \qquad X_{Fe}^{M2} = X_1 + X_3 = 1 - X_2 \tag{15}$$

$$X_{Mg}^{M1} = X_1 + X_2 = 1 - X_3 \qquad X_{Fe}^{M1} = X_3, \tag{16}$$

and we recognize that $X_1 + X_2 + X_3 = 1$ and that the anti-ordered component $(Mg)^{M2}(Fe)^{M1}Si_2O_6$ has $X_2 = X_3 = 1$ and $X_1 = -1$. We may then express the molar vibrational Gibbs energy using a Taylor expansion of second degree in terms of the independent mole fractions X_2 and X_3

$$\overline{G}^* = g_o + g_{X_2}X_2 + g_{X_3}X_3 + g_{X_2X_2}(X_2)^2 + g_{X_3X_3}(X_3)^2 + g_{X_2X_3}(X_2)(X_3). \tag{17}$$

The molar Gibbs energies of the endmembers $Mg_2Si_2O_6$ ($\overline{G}_2^o$) and $Fe_2Si_2O_6$ ($\overline{G}_3^o$), and the ordered and anti-ordered components $(Fe)^{M2}(Mg)^{M1}Si_2O_6$ ($\overline{G}_1^o$) and $(Mg)^{M2}(Fe)^{M1}Si_2O_6$ ($\overline{G}_4^o$) may be identified with these Taylor coefficients by setting X_2 and X_3 to their values in them:

$$\overline{G}_1^o = g_o \tag{18}$$

$$\overline{G}_2^o = g_o + g_{X_2} + g_{X_2X_2} \tag{19}$$

$$\overline{G}_3^o = g_o + g_{X_3} + g_{X_3X_3} \tag{20}$$

$$\overline{G}_4^o = g_o + g_{X_2} + g_{X_3} + g_{X_2X_2} + g_{X_3X_3} + g_{X_2X_3} \tag{21}$$

From these equation we readily discover that $g_{X_2X_3}$ is the reciprocal energy

$$g_{X_2X_3} = \Delta\overline{G}_{23}^o = \overline{G}_4^o + \overline{G}_1^o - \overline{G}_2^o - \overline{G}_3^o \tag{22}$$

which expresses the departure of the Gibbs energy of these endmembers and the ordered and anti-ordered components from coplanarity, with $\overline{G}_4^o$ lying above the plane defined by $\overline{G}_1^o$, $\overline{G}_2^o$, and $\overline{G}_3^o$ for positive $\Delta\overline{G}_{23}^o$. Finally, the $g_{X_2X_2}$ and $g_{X_3X_3}$ coefficients of the Taylor expansion may be equated with the negatives of the regular solution parameters for mixing of Fe and Mg on the M2 and M1 sites by considering mixing on the 1–2 and 1–3 joins and rewriting these expressions in the form:

$$\overline{G}^* = g_o(1 - X_2) + \left(g_o + g_{X_2} + g_{X_2X_2}\right)(X_2) - g_{X_2X_2}(X_2)(1 - X_2) \tag{23}$$

$$\overline{G}^* = g_o(1 - X_3) + \left(g_o + g_{X_3} + g_{X_3X_3}\right)(X_3) - g_{X_3X_3}(X_3)(1 - X_3), \tag{24}$$

and this leads to the following equation for the molar vibrational Gibbs energy

$$\begin{aligned}\overline{G}^* &= \overline{G}_1^\circ X_1 + \overline{G}_2^\circ X_2 + \overline{G}_3^\circ X_3 + W_{FeMg}^{M2}(X_2)(1 - X_2)\\ &+ W_{FeMg}^{M1}(X_3)(1 - X_3) + \Delta\overline{G}_{23}^\circ(X_2)(X_3)\end{aligned} \tag{25}$$

The molar configurational entropy, $\overline{S}^{IC}$, may be expressed by the relation:

$$\overline{S}^{IC} = -R \sum_{\alpha=M1,M2} \sum_{i=Mg,Fe} X_i^\alpha \ln X_i^\alpha, \tag{26}$$

$$= -R\left[X_{Mg}^{M1} \ln X_{Mg}^{M1} + X_{Fe}^{M1} \ln X_{Fe}^{M1} + X_{Mg}^{M2} \ln X_{Mg}^{M2} + X_{Fe}^{M2} \ln X_{Fe}^{M2}\right]. \tag{27}$$

Combining this expression with that for $\overline{G}^*$ with the relation

$$\overline{G} = \overline{G}^* - T\overline{S}^{IC} \tag{28}$$

we obtain the following expression for $\overline{G}$

$$\begin{aligned}\overline{G} = \overline{G}_1^\circ X_1 + \overline{G}_2^\circ X_2 + \overline{G}_3^\circ X_3 + W_{FeMg}^{M2}(X_2)(1 - X_2) + W_{FeMg}^{M1}(X_3)(1 - X_3)\\ + \Delta\overline{G}_{23}^\circ(X_2)(X_3)\\ + RT[X_2 \ln X_2 + (1 - X_2)\ln(1 - X_2) + X_3 \ln X_3 + (1 - X_3)\ln(1 - X_3)].\end{aligned} \tag{29}$$

According to the Darken equation the chemical potentials are given by

$$\mu_1 = \overline{G} + (1 - X_1)\left(\frac{\partial\overline{G}}{\partial X_1}\right)_{X_2/X_3}, \tag{30}$$

$$\frac{\partial X_2}{\partial X_1} = \frac{-X_2}{1 - X_1}, \quad \frac{\partial X_3}{\partial X_1} = \frac{-X_3}{1 - X_1}, \tag{31}$$

$$\mu_1 = \overline{G}_1^\text{o} + RTln\left(X_{Fe}^{M2}\right)\left(X_{Mg}^{M1}\right) + W_{FeMg}^{M2}\left(X_{Mg}^{M2}\right)^2 + W_{FeMg}^{M1}\left(X_{Fe}^{M1}\right)^2 - \Delta\overline{G}_{23}^\text{o}\left(X_{Mg}^{M2}\right)\left(X_{Fe}^{M1}\right), \tag{32}$$

$$\mu_2 = \overline{G} + (1 - X_2)\left(\frac{\partial\overline{G}}{\partial X_2}\right)_{X_1/X_3}, \tag{33}$$

$$\frac{\partial X_1}{\partial X_2} = \frac{-X_1}{1 - X_2}, \quad \frac{\partial X_3}{\partial X_2} = \frac{-X_3}{1 - X_2}, \tag{34}$$

$$\mu_2 = \overline{G}_2^o + RTln\left(X_{Mg}^{M2}\right)\left(X_{Mg}^{M1}\right) + W_{FeMg}^{M2}\left(X_{Fe}^{M2}\right)^2 + W_{FeMg}^{M1}\left(X_{Fe}^{M1}\right)^2 + \Delta\overline{G}_{23}^o\left(X_{Fe}^{M2}\right)\left(X_{Fe}^{M1}\right), \tag{35}$$

$$\mu_3 = \overline{G} + (1 - X_3)\left(\frac{\partial \overline{G}}{\partial X_3}\right)_{X_1/X_2}, \tag{36}$$

$$\frac{\partial X_2}{\partial X_3} = \frac{-X_2}{1 - X_3}, \quad \frac{\partial X_1}{\partial X_3} = \frac{-X_1}{1 - X_3}, \tag{37}$$

$$\mu_3 = \overline{G}_3^o + RTln\left(X_{Fe}^{M2}\right)\left(X_{Fe}^{M1}\right) + W_{FeMg}^{M2}\left(X_{Mg}^{M2}\right)^2 + W_{FeMg}^{M1}\left(X_{Mg}^{M1}\right)^2 + \Delta\overline{G}_{23}^o\left(X_{Mg}^{M2}\right)\left(X_{Mg}^{M1}\right). \tag{38}$$

Finally, to utilize the equations for chemical potentials, it is necessary to determine the equilibrium mole fractions X_{Fe}^{M2}, X_{Mg}^{M2}, X_{Fe}^{M1} and X_{Mg}^{M1} for any given Fe/(Fe + Mg) ratio and temperature. And these may be obtained by evaluating the condition of homogeneous equilibrium,

$$\left(\frac{\partial \overline{G}}{\partial X_1}\right)_{dX_2 = dX_3 = -\frac{1}{2}dX_1, T} = 0, \tag{39}$$

which yields the result

$$\begin{aligned} 0 = RTln\left(\frac{X_{Fe}^{M2}X_{Mg}^{M1}}{X_{Mg}^{M2}X_{Fe}^{M1}}\right) + \left(\overline{G}_1^o - \overline{G}_4^o\right) - W_{FeMg}^{M2}(1 - 2X_2) \\ - W_{FeMg}^{M1}(1 - 2X_3) + \Delta\overline{G}_{23}^o(1 - X_2 - X_3). \end{aligned} \tag{40}$$

Eq. (40) may be rearranged to

$$\begin{aligned} 0 = RTln\left(\frac{X_{Fe}^{M2}X_{Mg}^{M1}}{X_{Mg}^{M2}X_{Fe}^{M1}}\right) + \left(\overline{G}_1^o - \overline{G}_4^o\right) - W_{FeMg}^{M2}\left(2X_{Fe}^{M2} - 1\right) \\ - W_{FeMg}^{M1}\left(2X_{Mg}^{M1} - 1\right) + \Delta\overline{G}_{23}^o\left(X_{Fe}^{M2} + X_{Mg}^{M1} - 1\right), \end{aligned} \tag{41}$$

which is equivalent to eqs. (16) and (20) in [46, 47], and to eq. (3a) in [45] when is recognized that [45] uses a formula unit twice that employed here.

2.1.2 $(Cu, Ag)_{10}Fe_2Sb_4S_{13}$ *fahlores*

Because binary $(Cu, Ag)_6^{TRG}\left([Cu, Ag]_{2/3}Fe_{1/3}\right)_6^{TET}$ Sb_4S_{13} fahlores have two distinct crystallographic sites, (TRG) and (TET), on which Ag and Cu may order, but the sums of Cu and Ag differ on these sites, it will not be possible to readily model them as ternary reciprocal solutions when taking into account the ordering of Ag and Cu between trigonal-planar and tetrahedral sites in this $I\overline{4}3m$ derivative sphalerite structure with two formula units per unit cell (for example, Pauling and Newmann [51]).

Instead it is appropriate that we follow the approach of [28] and define the thermodynamic state of this solid solutions with a composition variable

$$X_4 = Ag/(Ag + Cu) \tag{42}$$

and an ordering variable that takes into account that the condition of homogeneous equilibrium for the distribution of Ag and Cu between trigonal-planar and tetrahedral metal sites due to the fact that the Ag-Cu exchange reaction is not a simple exchange reaction between these sites due the different multiplicities of Ag and Cu on them. Instead the Ag-Cu exchange reaction between trigonal-planar and tetrahedral metal sites must also involve the Ag and Cu endmembers for mass balance:

$$\begin{aligned} &\frac{1}{2}Cu_6^{TRG}\left(Ag_{2/3}, Fe_{1/3}\right)_6^{TET}Sb_4S_{13} + \frac{1}{10}Ag_{10}Fe_2Sb_4S_{13} = \\ &\frac{1}{2}Ag_6^{TRG}\left(Cu_{2/3}, Fe_{1/3}\right)_6^{TET}Sb_4S_{13} + \frac{1}{10}Cu_{10}Fe_2Sb_4S_{13}, \end{aligned} \tag{43}$$

as in other binary solutions where ordering occurs between sites of different multiplicities such as in Ag-Cu ordering in $(Ag,Cu)_{16}(Sb,As)_2S_{11}$ polybasite-pearceite and in Fe-Mg ordering in $(Mg,Fe)_7Si_8O_{22}OH_2$ cummintonite solutions (for example, [52, 53]). Accordingly, it is necessary to introduce an ordering variable into the formulation to make explicit provision for Ag-Cu ordering between trigonal-planar and tetrahedral metal sites. Following [28] we define this variable as

$$s = X_{Ag}^{TRG} - \frac{3}{2}X_{Ag}^{TET}, \tag{44}$$

$$= -\left(X_{Cu}^{TRG} - \frac{3}{2}X_{Cu}^{TET}\right), \tag{45}$$

a variable which has the value of zero when Ag and Cu are disordered or

$$\frac{X_{Ag}^{TRG}}{X_{Ag}^{TET}} = \frac{3}{2}, \tag{46}$$

and which has the value of +1 in the ordered component $(Ag)_6^{TRG}\left(Cu_{2/3}, Fe_{1/3}\right)_6^{TET}Sb_4S_{13}$ and −1 in the anti-ordered component $(Cu)_6^{TRG}\left(Ag_{2/3}, Fe_{1/3}\right)_6^{TET}Sb_4S_{13}$ (compare **Figure** 5).

Accordingly, the mole fractions of Cu and Ag on trigonal-planar and tetrahedral metal sites are now given by the expressions:

$$X_{Ag}^{TRG} = X_4 + \frac{2}{5}s \qquad X_{Cu}^{TRG} = 1 - X_4 - \frac{2}{5}s, \tag{47}$$

$$X_{Ag}^{TET} = \frac{2}{3}X_4 - \frac{2}{5}s \qquad X_{Cu}^{TET} = \frac{2}{3}(1 - X_4) + \frac{2}{5}s, \tag{48}$$

Following the methods outlined in [14] we may then write for the molar Gibbs energy of binary $(Cu, Ag)_{10}Fe_2Sb_2S_{13}$ fahlores

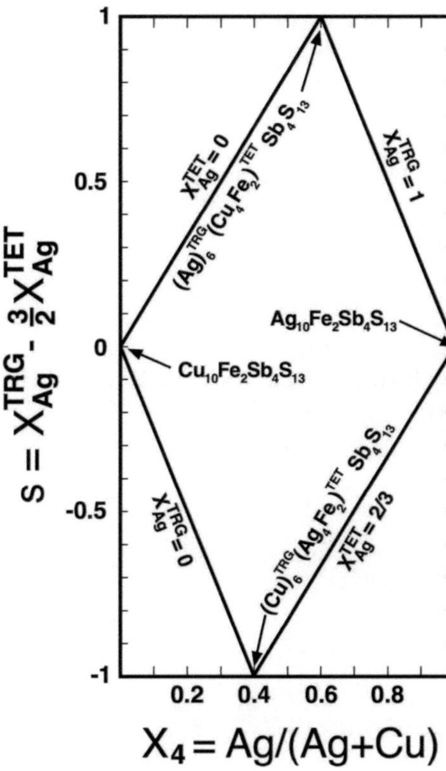

Figure 5.
Composition-ordering space for $(Cu, Ag)_{10}Fe_2Sb_4S_{13}$ *fahlores.*

$$\begin{aligned}\overline{G} = {} & \overline{G}_1^{\circ}X_1 + \overline{G}_4^{\circ}X_4 + \Delta\overline{G}_s^{*}s + \frac{1}{10}\left(\Delta\overline{G}_{4s}^{*} - 4W_{AgCu}^{TRG} + 6W_{AgCu}^{TET}\right)(2X_4 - 1)s \\ & + \left(\Delta\overline{G}_{4s}^{*} + W_{AgCu}^{TRG} + W_{AgCu}^{TET}\right)X_4(1 - X_4) + \frac{1}{25}\left(6\Delta\overline{G}_{4s}^{*} - 4W_{AgCu}^{TRG} - 9W_{AgCu}^{TET}\right)s^2 \\ & + RT[4\ln\left(\frac{3}{2}\right) + 6\left(X_4 + \frac{2}{5}s\right)\ln\left[X_4 + \frac{2}{5}s\right] + 6\left(1 - X_4 - \frac{2}{5}s\right)\ln\left[1 - X_4 - \frac{2}{5}s\right] \\ & + 6\left(\frac{2}{3}X_4 - \frac{2}{5}s\right)\ln\left[\frac{2}{3}X_4 - \frac{2}{5}s\right] + 6\left(\frac{2}{3}(1 - X_4) + \frac{2}{5}s\right)\ln\left[\frac{2}{3}(1 - X_4) + \frac{2}{5}s\right]],\end{aligned} \tag{49}$$

where $\Delta\overline{G}_s^{*}$ and $\Delta\overline{G}_{4s}^{*}$ are the Gibbs energies of ordering reaction (43) and of the reciprocal ordering reaction

$$\begin{aligned} & Cu_{10}Fe_2Sb_4S_{13} + Ag_{10}Fe_2Sb_4S_{13} = \\ & Ag_6^{TRG}(Cu_{2/3}, Fe_{1/3})_6^{TET}Sb_4S_{13} + Cu_6^{TRG}\left(Ag_{2/3}, Fe_{1/3}\right)_6^{TET}Sb_4S_{13},\end{aligned} \tag{50}$$

analogous to the reciprocal ordering reaction for Fe-Mg orthopyroxenes. From this expression the condition of homogeneous equilibrium for the ordering of Ag and Cu between trigonal-planar and tetrahedral metal sites may be readily evaluated by setting

$$\left(\frac{\partial\overline{G}}{\partial s}\right)_{T,X_4} = 0, \tag{51}$$

which results in the following expression:

$$0 = RT\ ln\frac{\left(X_4+\frac{2}{5}s\right)\left(\frac{2}{3}[1-X_4]+\frac{2}{5}s\right)}{\left(1-X_4-\frac{2}{5}s\right)\left(\frac{2}{3}X_4-\frac{2}{5}s\right)}+\frac{5}{12}\Delta\overline{G}_s^*$$
$$+\frac{1}{24}\left(\Delta\overline{G}_{4s}^*+6\ W_{AgCu}^{TET}-4\ W_{AgCu}^{TRG}\right)(2X_4-1)+\frac{1}{30}\left(6\ \Delta\overline{G}_{4s}^*-9\ W_{AgCu}^{TET}-4\ W_{AgCu}^{TRG}\right)(s). \tag{52}$$

Finally, chemical potentials of the endmember components $Cu_{10}Fe_2Sb_4S_{13}$ and $Ag_{10}Fe_2Sb_4S_{13}$ may be evaluated from an extended form of the Darken equation (for example, [44–47, 54, 55]):

$$\mu_i=\overline{G}+(1-X_i)\left(\frac{\partial G}{\partial X_i}\right)_{s,T}-s\left(\frac{\partial G}{\partial s}\right)_{X_4,T}, \tag{53}$$

which results in the following expressions for them:

$$\mu_1=\overline{G}_1^o+RTln\left[\left(1-X_4-\frac{2}{5}s\right)^6\left(\frac{2}{3}[1-X_4]+\frac{2}{5}s\right)^4\left(\frac{3}{2}\right)^4\right]$$
$$+\left(\Delta\overline{G}_{4s}^*+W_{AgCu}^{TRG}+W_{AgCu}^{TET}\right)(X_4)^2-\frac{1}{5}\left(\Delta\overline{G}_{4s}^*+6\ W_{AgCu}^{TET}-4\ W_{AgCu}^{TRG}\right)(X_4)s$$
$$-\frac{1}{25}\left(6\Delta\overline{G}_{4s}^*-4W_{AgCu}^{TRG}-9W_{AgCu}^{TET}\right)s^2, \tag{54}$$

and

$$\mu_4=\overline{G}_4^o+RTln\left[\left(X_4+\frac{2}{5}s\right)^6\left(\frac{2}{3}X_4-\frac{2}{5}s\right)^4\left(\frac{3}{2}\right)^4\right]+\left(\Delta\overline{G}_{4s}^*+W_{AgCu}^{TRG}+W_{AgCu}^{TET}\right)(1-X_4)^2$$
$$+\frac{1}{5}\left(\Delta\overline{G}_{4s}^*+6\ W_{AgCu}^{TET}-4\ W_{AgCu}^{TRG}\right)(1-X_4)s$$
$$-\frac{1}{25}\left(6\Delta\overline{G}_{4s}^*-4W_{AgCu}^{TRG}-9W_{AgCu}^{TET}\right)s^2. \tag{55}$$

What is readily apparent from examining Eqs. (49), (54) and (55) is the conundrum that the expression

$$\overline{G}=\mu_1X_1+\mu_4X_4 \tag{56}$$

does not appear to hold, as the thermodynamic parameter $\Delta\overline{G}_s^*$ appears in the equation for the Gibbs energy, but not in the expressions for the chemical potentials. However, it may readily be demonstrated by numerical analysis that Eq. (56) is indeed correct. The difficulties of eliminating s terms from such expressions which make explicit provisions for internal reactions and reducing them to expressions involving composition terms alone have been noted by Thompson [45] and others.

2.1.3 (Zn, Fe)S sphalerites

Sphalerites approximating the chemical formula (Zn,Fe)S are a common constituent of polymetallic base-metal sulfide deposits, occur in some iron and achondritic meteorites, and are a derivative of the diamond structure possessing a fcc lattice with $F\bar{4}3m$ space group symmetry. To aide deciphering the petrogenetic significance of sphalerite compositions and their variations in these occurrences, thermodynamic models of (Zn,Fe)S sphalerite have been constructed based on phase equilibrium, lattice dimensions, and activity-composition relations for it and coexisting sulfides and metals. Many of these models have been constructed on the premise that (Zn,Fe)S sphalerites may be described as an asymmetric simple solution:

$$\begin{aligned}\overline{G} \;=\; & \overline{G}^{o}_{ZnS}(1-X^{SPH}_{FeS}) + \overline{G}^{o}_{FeS}(X^{SPH}_{FeS}) + RT[(1-X^{SPH}_{FeS})\,ln\,(1-X^{SPH}_{FeS}) \\ & +X^{SPH}_{FeS}\,ln\,(X^{SPH}_{FeS})] \\ & +(X^{SPH}_{FeS})(1-X^{SPH}_{FeS})\left[W^{SPH}_{G\ ZnS}(X^{SPH}_{FeS}) + W^{SPH}_{G\ FeS}(1-X^{SPH}_{FeS})\right]\end{aligned} \tag{57}$$

with temperature and pressure dependent mixing parameters $W^{SPH}_{G\ ZnS}$ and $W^{SPH}_{G\ FeS}$ (for example, [42, 56–62]). Unfortunately these models are inadequate, because they fail to account for phase equilibrium features in detail (for example, [63]), do not account for the fact that the volumes of FeS-rich sphalerites synthesized at extremely low sulfur fugacities are dependent on the temperature of annealing (for example, **Figure 3** in [5]), produce excess entropies of mixing much greater than the maximum permitted by calorimetric data, 0.25–0.3 J/K-mol [48, 64], and do not account for the magnetic properties of Fe-substituted sphalerites (for example, [65–69]). Moreover, these data suggest abrupt changes in thermodynamic properties of (Zn,Fe)S sphalerites with increasing FeS (particularly by $X^{SPH}_{FeS} > 0.20 - 0.25$), and we cannot readily ascribe such changes to long-range ordering, as there is no X-ray evidence to support such a contention. Simple solution models such as that given by Eq. (57) (or by the quasi-chemical approximation or Landau expansions) do not provide an appropriate description, because, among other things, they do not take into account that next to nearest neighbor (*nnn*) interactions on the metallic sublattice may contribute substantially to the vibrational Gibbs energy, and that Fe-Zn clusters of various forms may modify the configurational entropy and energy relative to that due to random mixing of Fe and Zn on the metallic sublattice.

To attempt to address these deficiencies [48] developed a statistical mechanical, cluster variation method (CVM) model (for example, [70–72]) for (Zn,Fe)S sphalerite based on 13 and 6 metal cuboctahedral and octahedral basis clusters. In CVM models the configuration entropy (S^{IC}) of the solid solution is expressed in terms of configurational probabilities (or distribution numbers) of distinct atomic groups or clusters, with overlapping figures giving rise to a succession of smaller clusters or figures, ending in the lattice points individually (compare **Figure 6**). S^{IC} is typically given by an equation of the form:

$$S^{IC} = \sum_{a_i} \gamma(a_i) S^{a_i}, \tag{58}$$

where the summation is over the a_i basis clusters and overlap figures they form and $\gamma(a_i)$ are the entropy coefficients for each type of clusters a_i ($\gamma_{co} = \gamma_{oh} = 1$). The entropy attributed to each cluster a may be calculated from the formula

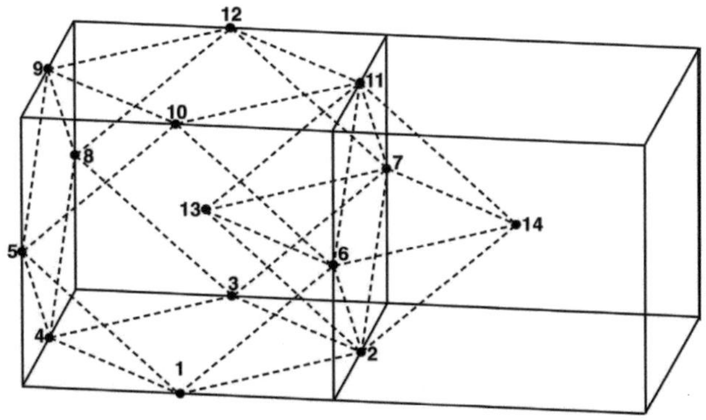

Figure 6.
*Cubeoctahedral (**co**) and octahedral (**oh**) basis clusters of the fcc metal lattice of (Zn,Fe)S sphalerite and overlap figures (clusters) formed from them: cube octahedron **co** [1-13], octahedron **oh** [14,11,7,6,2,13], double tetrahedron **dt** [1,5,4,2,6,13], quadrilateral pyramid **py** [6,2,7,11,13], regular tetrahedron **td** [1,4,5,13], irregular tetrahedron **itd** [1,13,5,6], square **sq** [1,2,3,4], equilateral triangle **tr** [1,4,5], right angular triangle **rtr** [1,3,13], first nn pair **nn** [1,2]. second nn pair **nnn** [6,7] and point cluster. From Balabin and Sack ([48], fig. 2).*

$$S^a = -Nk_B \sum_l \alpha_l^a p_l^a \ln p_l^a, \tag{59}$$

where N is the number of lattice points, k_B is Boltmann's constant, and p_l^a and α_l^a are the probability and multiplicity of state l of cluster a. For the approximation of Balabin and Sack [48] these equations produce the following equation for the configurational entropy

$$S^{IC} = S^{\mathbf{co}} + S^{\mathbf{oh}} - 6S^{\mathbf{dt}} + 6S^{\mathbf{td}} - 6S^{\mathbf{py}} + 12S^{\mathbf{itd}} - 8S^{\mathbf{tr}}, \tag{60}$$

where **co**, **oh**, **dt**, **td**, **py**, **itd**, and **tr** refer respectively to the cube octahedron, octahedron, double tetrahedron, regular tetrahedron, quadrilateral pyramid, irregular tetrahedron, and equilateral triangle figures illustrated in **Figure 6**.

To describe the volume [48] developed a bond length approximation based on the partition of the lattice space into fully connected octahedra and tetrahedra, using the well known formula expressing tetrahedral volume as a function of cell edge and Zn-Zn, Fe-Fe and Zn-Fe bond lengths (d_{Zn-Zn},d_{Zn-Fe},d_{Fe-Fe}), and formulae for the octahedral volumes for the 10 different configurations of Fe and Zn metals in them given in their Table 3 expressed in these bond lengths. From these formula the configurational volume may be expressed as:

$$V = N\sum_{i=1}^{10} \alpha_i^{\mathbf{oh}} V_i^{\mathbf{oh}} p_i^{\mathbf{oh}} + 2N\sum_{j=1}^{5} \alpha_j^{\mathbf{td}} V_j^{\mathbf{td}} p_j^{\mathbf{td}}, \tag{61}$$

where $V_i^{\mathbf{oh}}$ and $V_j^{\mathbf{td}}$ are the configurational volumes of the i and j configurations of the octahedron and tetrahedron, respectively. And finally, Balabin and Sack [48] calculated the internal energy based on three different assumptions allowing for both pair and multibodied interactions of increasing size: (1) pair interactions between nearest neighbor (**nn**) and next nearest neighbor (**nnn**) pairs, (2) next nearest neighbor (**nnn**) pair interactions and interactions between (**nn**) pairs in equilateral triangles (**tr**), and (3) interactions within equilateral triangles (**tr**) and centered squares (**csq**). In the first case the internal energy per lattice site (E) may be given by the formula

$$E = E_{Zn}(1 - X_{Fe}) + E_{Fe}(X_{Fe}) - 6w^{\mathbf{nn}}p_2^{\mathbf{nn}} - 3w^{\mathbf{nnn}}p_2^{\mathbf{nnn}}, \tag{62}$$

where w^{nn} and w^{nnn} are the interchange energies for Zn and Fe atoms defined by the relations:

$$w^{nn} = \epsilon_1^{\mathbf{nn}} + \epsilon_3^{\mathbf{nn}} - 2\epsilon_2^{\mathbf{nn}} \tag{63}$$

and

$$w^{nnn} = \epsilon_1^{\mathbf{nnn}} + \epsilon_3^{\mathbf{nnn}} - 2\epsilon_2^{\mathbf{nnn}}, \tag{64}$$

and where $\epsilon_1^{\mathbf{nn}}$, $\epsilon_2^{\mathbf{nn}}$, and $\epsilon_3^{\mathbf{nn}}$ and $\epsilon_1^{\mathbf{nnn}}$, $\varepsilon_2^{\mathbf{nnn}}$, and $\varepsilon_3^{\mathbf{nnn}}$ are the energies of Zn-Zn, Fe-Zn and Fe-Fe *nn* and *nnn* bonds, respectively. To define the internal energy per lattice site in the second approximation three parameters are required, the $w^{\mathbf{nnn}}$ interchange energy defined above, and the parameters

$$w_1^{\mathbf{tr}} = \epsilon_1^{\mathbf{tr}} - 3\epsilon_3^{\mathbf{tr}} + 2\epsilon_4^{\mathbf{tr}}, \tag{65}$$

and

$$w_2^{\mathbf{tr}} = \epsilon_2^{\mathbf{tr}} - 2\epsilon_3^{\mathbf{tr}} + \epsilon_4^{\mathbf{tr}}, \tag{66}$$

where $\epsilon_1^{\mathbf{tr}}$, $\epsilon_2^{\mathbf{tr}}$, and $\epsilon_3^{\mathbf{tr}}$ and, $\epsilon_4^{\mathbf{tr}}$ are the energies of the Zn-Zn-Zn, Zn-Zn-Fe, Zn-Fe-Fe, and Fe-Fe-Fe equilateral triangles. And the internal energy per lattice site becomes

$$E \;=\; E_{Zn}(1 - X_{Fe}) + E_{Fe}(X_{Fe}) - \frac{8}{3}w_1^{\mathbf{tr}}\left(2p_2^{\mathbf{tr}} + p_3^{\mathbf{tr}}\right) + 8w_2^{\mathbf{tr}}p_2^{\mathbf{tr}} - \frac{3}{2}w^{\mathbf{nnn}}p_2^{\mathbf{nnn}}. \tag{67}$$

Finally, 10 parameters are required to describe the internal energy when both the equilateral triangle (**tr**) and centered square (**csq**) are chosen as the energy embodying multimetal clusters and the appropriate formulae are given by Balabin and Sack [48].

At a given X_{FeS}^{SPH}, temperature and pressure the equilibrium state is determined by minimizing the Gibbs energy with respect to variations in the 283 configurational probabilities that result from the 288 and 10 independent configurations of the cube octahedron (**co**) and octahedron (**oh**), respectively, when allowance is made for 15 relationships provided by consistency requirements, compositional closure, the normalizing conditions that

$$\sum_i^{288} \alpha_i^{\mathbf{co}} p_i^{\mathbf{co}} = \sum_i^{10} \alpha_i^{\mathbf{oh}} p_i^{\mathbf{oh}} \;=\; 1, \tag{68}$$

and the requirements that all $p_i^{\mathbf{co}}$ and $p_i^{\mathbf{oh}} > 0$. The consistency relations arise from the fact that the probabilities of the configurations of subclusters are completely determined by the probabilities of the configurations of the clusters from which they are derived. For each subcluster that may be derived from q basis and other clusters, there are q–1 linear relationships between the probabilities of the states of the basis components, and these were implemented by "brute force" during Gibbs energy minimization. The explicit formulation and implementation of the linear equations relating different cluster configurations, a development of the principles formulated by Hijmans and de Boir [71], contrasts with the use of so-called correlation polynomials normally employed as independent variables in CVM calculations (for example, [73–76]). It affords considerable improvements in computational efficiencies, and the calculations may be readily performed at very low concentrations ($>10^{-18} - 10^{-20}$ mol %), whereas computational difficulties may be encountered at much higher concentrations (<1 mol %), when configurational polynomials are employed as independent variables.

In their analysis [48] discovered that consideration of the energies of at least one many bodied interaction (**tr**) was necessary to recover the salient features of the active constraints, their model employing three empirical interaction parameters for next to nearest neighbor (**nnn**) pair interactions and for three-bodied interactions associated with nearest neighbor interactions (**nn**) in equilateral triangles (**tr**) accounting for all experimental data on Fe-solubility in sphalerite coexisting with pyrrhotite and pyrite and with pyrrhotite and iron metal within experimental uncertainty, predicting positive deviations from ideality consistent with those obtained by Fleet [77] by Gibbs-Duhem integration of the data of Barton and Toulmin [5] at 850° C, and having excess entropies permitted by the calorimetric data of Pankratz and King [64]. The model also correctly predicts that the cell edge of quenched Fe-sphalerites decreases with synthesis temperature, but with significantly less amplitude than suggested by the data of Barton and Toulmin [5] (their **Figure 3**). But like the more sophisticated (**tr**) + (**csq**) model with 12 adjustable parameters (10 energy parameters and bond lengths d_{Zn-Fe} and d_{Fe-Fe}), it also suggests that long-range ordering to lower symmetry superstructures, in addition to short-range ordering, may complicate phase relations at the lower temperatures, a prediction that appears to accord with widespread "metastable" phenomena associated with lower temperature (<500°C) experiments (for example, [56, 78, 79]) and may provide potential insights into sphalerite/wurzite polytypes (for example, [80, 81]) and repetitive sphalerite zoning (for example, [6, 82, 83]).

The inference that clusters of Fe atoms of increasing size are stabilized with increasing X_{FeS}^{SPH} has been shown to be consistent with Raman spectra (for example, [84]), Mössbauer spectroscopy (for example, [69]), and magnetic susceptibility [67, 69, 85–87]. Osachii and Gorbaty [84] conclude that compositional changes in Raman spectra indicate sphalerite with $X_{FeS}^{SPH} > 0.15$ undergo polymerization of (**nn**) pairs in clusters of increasing size, suggesting a structure change or percolation transition occurs with X_{FeS}^{SPH} between 0.15 and 0.25. Di Benedetto et al. [69] demonstrate that, for synthetic and natural sphalerites with $0.005 \leq X_{FeS}^{SPH} \leq 0.25$, the self affinity of Fe ions favored by superexchange p–d interactions between Fe and S (for example, [67, 69]) stabilizes the formation of clusters, even in Fe-dilute sphalerite. Furthermore, the first principles *ab-initio* calculations of Wright and Gale [88] indicate that Fe^{2+}-Fe^{2+} pairs are stabilized at low X_{FeS}^{SPH}, there is no energetic impediment to cluster formation at high X_{FeS}^{SPH}, and that IR spectra of Fe-sphalerites (for example, [89]) may be insensitive to clustering. Finally, Balabin and Sack [48] concluded that short-range-ordering in sphalerite is driven by a tendency to minimize the concentration of identical atoms in the second (**nnn**) rather than the first (**nn**) coordination sphere in the cationic sublattice ($w^{nnn} >> w_1^{tr}, w_2^{tr}$).

For practical use (that is, avoid minimizing the Gibbs energy with respect to 283 cluster configuration probabilities at each temperature, pressure and Fe-content of sphalerite of interest), [48] fitted their results to a fourth order Guggenheim polynomial (for example, [90]):

$$\overline{G}^{EX} = X_{FeS}^{SPH}\left(1 - X_{FeS}^{SPH}\right)\left[A_o + A_1\left(1 - 2X_{FeS}^{SPH}\right) + A_2\left(1 - 2X_{FeS}^{SPH}\right)^2 + A_3\left(1 - 2X_{FeS}^{SPH}\right)^3 + A_4\left(1 - 2X_{FeS}^{SPH}\right)^4\right] \quad (69)$$

with the coefficients A_i expressed as polynomials in T (K) ([48], p. 937):

$$A_o = 3464.954 + 4.9152\,T - 4.0522\left(T^2/10^3\right) + 1.20078\left(T^3/10^6\right).$$

$$A_1 = -4864.33 + 3.9523\,T - 3.50847\left(T^2/10^3\right) + 1.08473\left(T^3/10^6\right).$$

$$A_2 = 350.802 + 1.5184\,T - 1.57187\left(T^2/10^3\right) + 0.52146\left(T^3/10^6\right).$$

$$A_3 = 2326.79 - 5.6524\,T + 5.10041\left(T^2/10^3\right) - 1.59157\left(T^3/10^6\right).$$

$$A_4 = 3541.39 - 8.36153\,T + 7.44021\left(T^2/10^3\right) - 2.30296\left(T^3/10^6\right).$$

It is this calibration that we will employ in our subsequent discussions and calculations. For the volume of sphalerite we may employ the expression given by Chareev et al. [91] which is quadratic in X_{FeS} and linear in temperature over 273.15–723.15 K:

$$\overline{V} = \overline{V}_{ZnS}^{o}\left(1 - X_{FeS}\right) + \overline{V}_{FeS}^{o}X_{FeS} + W_v^{EX}\left(X_{FeS}\right)\left(1 - X_{FeS}\right) + \alpha_T(T - 298.15K), \quad (70)$$

where $\overline{V}_{ZnS}^{o} = 23.8374$, $\overline{V}_{FeS}^{o} = 23.9857$, $W_v^{EX} = 0.6186$, and $\alpha_T = 4.7405 \times 10^{-12}$ cc, and where $\overline{V}_{FeS}^{o}$ is between that determined for metastable FeS sphalerite by Refs. [92, 93]. Chareev et al. [91] make no provision for the volumes of quenched Fe-sphalerites to decrease with synthesis temperature, as suggested by the data of Barton and Toulmin [5]. Chareev et al. [91] synthesized Fe-sphalerites using both gas transport and molten salt methods, as opposed to the dry synthesis in silica glass ampoule techniques employed by Refs. [5, 91] and others (for example, [94]) have suggested their discrepancies with the [5] data may reflect reactions between their charges and ampoules at temperatures above 723.15 K. This inference might well account for the fact that [48] were unable to reproduce the amplitude of the dependence of volumes of FeS-rich sphalerites on temperature of annealing at the higher temperatures suggested by the data of Barton and Toulmin [5] (their **Figure 3**).

2.2 Quaternary $(Ag,Cu)_{10}(Fe,Zn)_2(Sb,As)_4S_{13}$ fahlores

Fahlores are the principal source of Ag to the mining industry and are a common constituent of Ag-Pb-Zn sulfide ore deposits where they may exhibit pronounced compositional zoning in time and space (for example, [2, 8, 10]). Studies of natural fahlores in such ores have documented that the vast majority of them approximate the chemical formula $(Cu,Ag)_{10}(Fe,Zn)_2(Sb,As)_4S_{13}$ and may have minor, and occasionally major, amounts of Mn, Cd, and Hg substituting for Fe, Se substituting for S, Te substituting for Sb and As (typically through the coupled substitutions CuTe for Fe (Sb,As) or □Te for Cu(Sb,As)), and $CuFe^{3+}$ substituting for $2Fe^{2+}$, in addition to a plethora of other minor substitutions (for example, [19, 30]). The stability of (Cu, Ag)$_{10}$(Fe,Zn)$_2$(Sb,As)$_4$S$_{13}$ compositions may be attributed to the fact that they are formally charge balanced and have the stable band structure associated with a full 52nd brillioun zone consisting of 208 valence electrons in the unit cell of two formula units (for example, [95–97]). In addition, a thermodynamic model for $(Cu,Ag)_{10}(Fe,Zn)_2(Sb,As)_4S_{13}$ fahlores has been developed (for example, [14, 19, 23, 28]) that is consistent with experimental and petrological constraints (for example, [4, 15, 16, 21, 23, 25, 28, 98–107]), and successsfully predicted the temperatures and compositions of binodal pairs of miscibility gaps in $(Cu,Ag)_{10}(Fe,Zn)_2Sb_4S_{13}$ fahlores that were subsequently confirmed [21, 24, 31–33]. The fact that this model has successfully predicted the existence and character of these gaps, and is predicated on the conclusion derived from the thermodynamic analysis of Fe-Zn exchange equilibrium between sphalerite and fahlore [28] that Ag undergoes a change in site preference from trigonal-planar (triangular) to tetrahedral sites well before six Ag replace Cu in the structural formula unit,

$$(Cu,Ag)_6^{TRG}\left[(Cu,Ag)_{2/3}(Zn,Fe)_{1/3}\right]_6^{TET}(As,Sb)_4^{SM}S_{13}, \tag{71}$$

stands in stark contrast to the conventional "wisdom" as summarized by this statement from Makovicky ([30], p. 54): "Substitution by Ag is universally assumed to start in the triangular sites and spill over in the tetrahedral sites only when the former are (nearly) filled." But this conventional "wisdom" is also inconsistent with reversed experimental brackets on Ag-Cu exchange equilibria (for example, [100, 101]), as well as the existence of the miscibility gaps documented for $(Cu,Ag)_{10}(Fe,Zn)_2Sb_4S_{13}$ fahlores. Furthermore, [18, 22] have demonstrated that this thermodynamic model for $(Cu,Ag)_{10}(Fe,Zn)_2Sb_4S_{13}$ fahlores coupled with those for coexisting sulfide, silicates, oxides and hydrothermal fluids result in predicted hydrothermal fluid compostions consistent with the mineral assemblages, mineral compositions, order of crystallization and other chemical and mineralogical features of ores from the Coeur d'Alene mining district (ID, USA) and Keno Hill mining district (Yukon, CA), when the primary compositions of these minerals are properly reconstructed (for example, [4, 15, 16, 21]). Finally, Sack [19] produced a preliminary calibration for activity-composition relations in $(Cu,Ag)_{10}(Fe,Zn)_2(Sb,As)_4S_{13}$ fahlores consistent with constraints on phase relations provided by the composition data of Ixer and Stanley [108] for fahlores from the Ag-Pb-Zn ores of the Hope mine on little Sark Island, Channel Islands, Great Britain. The resulting calibration of the model is consistent with the inference that As stabilizes Ag in trigonal-planar sites to produce Ag-Cu ordering dependencies on Ag/Cu ratio in As fahlores more like those incorrectly inferred for Sb fahlores according to conventional "wisdom". This then completes the preliminary calibration of the thermodynamic model with the important exception that the Gibbs energy of at least one As vertex of the $(Cu,Ag)_{10}(Fe,Zn)_2(Sb,As)_4S_{13}$ fahlore cube must be constrained, as only one of these endmember As vertices is linearly independent in the current formulation of this model.

Accordingly, we will focus our attention on this quaternary fahlore model subsystem and associated sulfides in our subsequent analysis. To do so we expand the treatment of binary $(Cu,Ag)_{10}Fe_2Sb_4S_{13}$ fahlores given above to make explicit provision for the substitutions of Zn for Fe and As for Sb employing the composition variables

$$X_2 = 3X_{Zn}^{TET} = 1 - 3X_{Fe}^{TET} \tag{72}$$

and

$$X_3 = X_{As}^{SM} = 1 - X_{Sb}^{SM}, \tag{73}$$

retaining the definitions of X_4 and s given by Eqs. (42) and (44). Following the methods outlined in [14] and above, these additions result in the following equations for the molar Gibbs energy and condition of homogeneous equilibrium:

$$\begin{aligned}\overline{G} = {} & \overline{G}_1^{\circ}X_1 + \overline{G}_2^{\circ}X_2 + \overline{G}_3^{\circ}X_3 + \overline{G}_4^{\circ}X_4 + \Delta\overline{G}_s^{*}s + \Delta\overline{G}_{23}^{\circ}X_2X_3 + \Delta G_{24}^{\circ}X_2X_4 \\ & + \Delta\overline{G}_{34}^{\circ}X_3X_4 + W_{FeZn}^{TET}X_2(1-X_2) + W_{AsSb}^{SM}X_3(1-X_3) + \Delta\overline{G}_{2s}^{*}X_2s \\ & + \Delta\overline{G}_{3s}^{*}X_3s + \frac{1}{10}\left(\Delta\overline{G}_{4s}^{*} - 4W_{AgCu}^{TRG} + 6W_{AgCu}^{TET}\right)(2X_4-1)s \\ & + \left(\Delta\overline{G}_{4s}^{*} + W_{AgCu}^{TRG} + W_{AgCu}^{TET}\right)X_4(1-X_4) + \frac{1}{25}\left(6\Delta\overline{G}_{4s}^{*} - 4W_{AgCu}^{TRG} - 9W_{AgCu}^{TET}\right)s^2 \\ & + RT[\ln\left[\left(\frac{3}{2}\right)^4 3^2\right] + 2X_2\ln\left[\frac{1}{3}X_2\right] + 2(1-X_2)\ln\left[\frac{1}{3}(1-X_2)\right] + 4X_3\ln[X_3] \\ & + 4(1-X_3)\ln[1-X_3] + 6\left(X_4+\frac{2}{5}s\right)\ln\left[X_4+\frac{2}{5}s\right] + 6\left(1-X_4-\frac{2}{5}s\right)\ln\left[1-X_4-\frac{2}{5}s\right] \\ & + 6\left(\frac{2}{3}X_4-\frac{2}{5}s\right)\ln\left[\frac{2}{3}X_4-\frac{2}{5}s\right] + 6\left(\frac{2}{3}(1-X_4)+\frac{2}{5}s\right)\ln\left[\frac{2}{3}(1-X_4)+\frac{2}{5}s\right]]\end{aligned} \tag{74}$$

and

$$\begin{aligned}0 = {} & RT\ ln\frac{\left(X_4+\frac{2}{5}s\right)\left(\frac{2}{3}[1-X_4]+\frac{2}{5}s\right)}{\left(1-X_4-\frac{2}{5}s\right)\left(\frac{2}{3}X_4-\frac{2}{5}s\right)} + \frac{5}{12}\left(\Delta\overline{G}_s^{*} + \Delta\overline{G}_{2s}^{*}(X_2) + \Delta\overline{G}_{3s}^{*}(X_3)\right) \\ & + \frac{1}{24}\left(\Delta\overline{G}_{4s}^{*} + 6\ W_{AgCu}^{TET} - 4\ W_{AgCu}^{TRG}\right)(2X_4-1) + \frac{1}{30}\left(6\ \Delta\overline{G}_{4s}^{*} - 9\ W_{AgCu}^{TET} - 4\ W_{AgCu}^{TRG}\right)(s),\end{aligned} \tag{75}$$

where $\Delta\overline{G}_{2s}^{*}$ and $\Delta\overline{G}_{3s}^{*}$ are the reciprocal ordering reactions

$$\begin{aligned}& \frac{1}{2}Cu_6^{TRG}\left(Ag_{2/3},Zn_{1/3}\right)_6^{TET}Sb_4S_{13} + \frac{1}{10}Cu_{10}Fe_2Sb_4S_{13} + \frac{1}{2}Ag_6^{TRG}\left(Cu_{2/3},Fe_{1/3}\right)_6^{TET}Sb_4S_{13} \\ & + \frac{1}{10}Ag_{10}Zn_2Sb_4S_{13} = \frac{1}{2}Ag_6^{TRG}\left(Cu_{2/3},Zn_{1/3}\right)_6^{TET}Sb_4S_{13} + \frac{1}{10}Ag_{10}Fe_2Sb_4S_{13} \\ & + \frac{1}{2}Cu_6^{TRG}\left(Ag_{2/3},Fe_{1/3}\right)_6^{TET}Sb_4S_{13} + \frac{1}{10}Cu_{10}Zn_2Sb_4S_{13}\end{aligned} \tag{76}$$

and

$$\begin{aligned}& \frac{1}{2}Cu_6^{TRG}\left(Ag_{2/3},Fe_{1/3}\right)_6^{TET}As_4S_{13} + \frac{1}{10}Cu_{10}Fe_2Sb_4S_{13} + \frac{1}{2}Ag_6^{TRG}\left(Cu_{2/3},Fe_{1/3}\right)_6^{TET}Sb_4S_{13} \\ & + \frac{1}{10}Ag_{10}Fe_2As_4S_{13} = \frac{1}{2}Ag_6^{TRG}\left(Cu_{2/3},Fe_{1/3}\right)_6^{TET}As_4S_{13} + \frac{1}{10}Ag_{10}Fe_2Sb_4S_{13} \\ & + \frac{1}{2}Cu_6^{TRG}\left(Ag_{2/3},Fe_{1/3}\right)_6^{TET}Sb_4S_{13} + \frac{1}{10}Cu_{10}Fe_2As_4S_{13}.\end{aligned} \tag{77}$$

Chemical potentials of endmember vertices of composition-ordering space are readily derived from the extended form of the Darken equation which makes explicit provision for Ag-Cu ordering:

$$\mu_j = \overline{G} + \sum_{i=1}^{4} n_{ij}(1 - X_i)\left(\frac{\partial \overline{G}}{\partial X_i}\right)_{T,X_k/X_l,s} + (s_j - s)\left(\frac{\partial \overline{G}}{\partial s}\right)_{T,X_2,X_3,X_4}. \tag{78}$$

In this equation the n_{ij} coefficients are the mole fractions of the *i* components ($Cu_{10}Fe_2Sb_4S_{13}$, $Cu_{10}Zn_2Sb_4S_{13}$, $Cu_{10}Fe_2Sb_4S_{13}$, and $Ag_{10}Fe_2Sb_4S_{13}$) in the *j* vertex of composition–ordering space of interest, where

$$X_1 = 1 - X_2 - X_3 - X_4, \tag{79}$$

and s_j is the value of s (−1, 0 or + 1), in this vertex. Explicit expressions for the chemical potentials of endmember *j* vertices with $s_j = 0$ (that is, the vertices of the fahlore composition cube in **Figure 7**: $Cu_{10}Fe_2Sb_4S_{13}$, $Cu_{10}Zn_2Sb_4S_{13}$, $Cu_{10}Fe_2Sb_4S_{13}$, $Ag_{10}Fe_2Sb_4S_{13}$, $Cu_{10}Zn_2As_4S_{13}$, $Ag_{10}Zn_2Sb_4S_{13}$, $Ag_{10}Fe_2As_4S_{13}$, and $Ag_{10}Zn_2As_4S_{13}$) are given in ([109], their table 1) and corresponding expressions for the chemical potentials of endmember *j* component vertices with s = −1 and + 1 are readily derived from Eq. (78) as detailed for vertex $Ag_6(Cu_4,Fe_2)Sb_4S_{13}$ in **Table 1**.[1]

---

[1] We want an expression for $\mu_j = \mu_{Ag_6^{TRG}Cu_4^{TET}Fe_2^{TET}Sb_4S_{13}}$. To make life easy we will assume all regular solution parameters (W^{TET}_{FeZn}, W^{TRG}_{AgCu}, W^{TET}_{AgCu}, W^{SM}_{AsSb}) and reciprocal parameters ($\Delta\overline{G}^{o}_{23}$, $\Delta\overline{G}^{o}_{24}$, $\Delta\overline{G}^{o}_{34}$) are zero. This reduces Eq. (74) to

$$\overline{G} = \overline{G}^{\circ}_1 X_1 + \overline{G}^{\circ}_2 X_2 + \overline{G}^{\circ}_3 X_3 + \overline{G}^{\circ}_4 X_4 + \Delta\overline{G}^{*}_s s + \Delta\overline{G}^{*}_{2s}(X_2)s + \Delta\overline{G}^{*}_{3s}(X_3)s + \Delta\overline{G}^{*}_{4s}\left[\frac{1}{10}(2X_4 - 1)s + X_4(1 - X_4) + \frac{6}{25}s^2\right]$$
$$+RT[\ln\left[\left(\frac{3}{2}\right)^4 3^2\right] + 2X_2\ln\left[\frac{1}{3}X_2\right] + 2(1 - X_2)\ln\left[\frac{1}{3}(1 - X_2)\right] + 4X_3\ln[X_3]$$
$$+4(1 - X_3)\ln[1 - X_3] + 6\left(X_4 + \frac{2}{5}s\right)\ln\left[X_4 + \frac{2}{5}s\right] + 6\left(1 - X_4 - \frac{2}{5}s\right)\ln\left[1 - X_4 - \frac{2}{5}s\right]$$
$$+6\left(\frac{2}{3}X_4 - \frac{2}{5}s\right)\ln\left[\frac{2}{3}X_4 - \frac{2}{5}s\right] + 6\left(\frac{2}{3}(1 - X_4) + \frac{2}{5}s\right)\ln\left[\frac{2}{3}(1 - X_4) + \frac{2}{5}s\right]].$$

We can further simplify the problem by combining the terms $RT\ln(3^2)$, $RT(2X_2\ln[\frac{1}{3}X_2])$ and $RT(2(1 - X_2)\ln[\frac{1}{3}(1 - X_2)])$:

$$\overline{G} = \overline{G}^{\circ}_1 X_1 + \overline{G}^{\circ}_2 X_2 + \overline{G}^{\circ}_3 X_3 + \overline{G}^{\circ}_4 X_4 + \Delta\overline{G}^{*}_s s + \Delta\overline{G}^{*}_{2s}(X_2)s + \Delta\overline{G}^{*}_{3s}(X_3)s + \Delta\overline{G}^{*}_{4s}\left[\frac{1}{10}(2X_4 - 1)s + X_4(1 - X_4) + \frac{6}{25}s^2\right]$$
$$+RT[\ln\left[\left(\frac{3}{2}\right)^4\right] + 2X_2\ln[X_2] + 2(1 - X_2)\ln[(1 - X_2)] + 4X_3\ln[X_3]$$
$$+4(1 - X_3)\ln[1 - X_3] + 6\left(X_4 + \frac{2}{5}s\right)\ln\left[X_4 + \frac{2}{5}s\right] + 6\left(1 - X_4 - \frac{2}{5}s\right)\ln\left[1 - X_4 - \frac{2}{5}s\right]$$
$$+6\left(\frac{2}{3}X_4 - \frac{2}{5}s\right)\ln\left[\frac{2}{3}X_4 - \frac{2}{5}s\right] + 6\left(\frac{2}{3}(1 - X_4) + \frac{2}{5}s\right)\ln\left[\frac{2}{3}(1 - X_4) + \frac{2}{5}s\right]].$$

For this j component $Ag_6^{TRG}Cu_4^{TET}Fe_2^{TET}Sb_4S_{13}$ we have n_{ij}'s of $n_{1j} = \frac{2}{5}$ and $n_{4j} = \frac{3}{5}$ and $s_j = +1$. Thus equation (78) reduces to

$$\mu_j = \overline{G} + \tfrac{2}{5}(1 - X_1)\left(\tfrac{\partial \overline{G}}{\partial X_1}\right)_{T,X_2/X_3,X_2/X_4,s} + \tfrac{3}{5}(1 - X_4)\left(\tfrac{\partial \overline{G}}{\partial X_4}\right)_{T,X_1/X_2,X_1/X_3,s} + (1 - s)\left(\tfrac{\partial \overline{G}}{\partial s}\right)_{T,X_2,X_3,X_4}.$$

This may then be readily shown to produce the result:

$$\mu_j = \frac{2}{5}\overline{G}^{\circ}_1 + \frac{3}{5}\overline{G}^{\circ}_4 + \Delta\overline{G}^{*}_s + \Delta\overline{G}^{*}_{2s}(X_2)(1 - s) + \Delta\overline{G}^{*}_{3s}(X_3)(1 - s)$$
$$+\frac{1}{10}\Delta\overline{G}^{*}_{4s}\left(5 - 10X_4 + 10X_4^2 + 6s - 2X_4 s - \frac{12}{5}s^2\right)$$
$$+RT\left[6\ln\left[X_4 + \frac{2}{5}s\right] + 4\ln\left[\left(\frac{3}{2}\right)\right]\left[\frac{2}{3}(1 - X_4) + \frac{2}{5}s\right] + 2\ln[1 - X_2] + 4\ln[1 - X_3]\right].$$

Which reduces to

$$\mu_j = \frac{2}{5}\overline{G}^{\circ}_1 + \frac{3}{5}\overline{G}^{\circ}_4 + \Delta\overline{G}^{*}_s + \frac{1}{2}\Delta\overline{G}^{*}_{4s}$$
$$+RT[6\ln[1] + 4\ln[1] + 2\ln[1] + 4\ln[1]].$$

Vertex	X_1	X_2	X_3	X_4	s_j
$Cu_{10}Fe_2Sb_4S_{13}$	1	0	0	0	0
$Cu_{10}Zn_2Sb_4S_{13}$	0	1	0	0	0
$Cu_{10}Fe_2As_4S_{13}$	0	0	1	0	0
$Ag_{10}Fe_2Sb_4S_{13}$	0	0	0	1	0
$Cu_{10}Zn_2As_4S_{13}$	−1	1	1	0	0
$Ag_{10}Zn_2Sb_4S_{13}$	−1	1	0	1	0
$Ag_{10}Fe_2As_4S_{13}$	−1	0	1	1	0
$Ag_{10}Zn_2As_4S_{13}$	−2	1	1	1	0
$Ag_6(Cu_4,Fe_2)Sb_4S_{13}^{*}$	$+\frac{2}{5}$	0	0	$+\frac{3}{5}$	+1
$Cu_6(Ag_4,Fe_2)Sb_4S_{13}^{*}$	$+\frac{3}{5}$	0	0	$+\frac{2}{5}$	−1
$Ag_6(Cu_4,Zn_2)Sb_4S_{13}^{*}$	$-\frac{3}{5}$	1	0	$+\frac{3}{5}$	+1
$Cu_6(Ag_4,Zn_2)Sb_4S_{13}^{*}$	$-\frac{2}{5}$	1	0	$+\frac{2}{5}$	−1
$Ag_6(Cu_4,Fe_2)As_4S_{13}^{*}$	$-\frac{3}{5}$	0	1	$+\frac{3}{5}$	+1
$Cu_6(Ag_4,Fe_2)As_4S_{13}^{*}$	$-\frac{2}{5}$	0	1	$+\frac{2}{5}$	−1
$Ag_6(Cu_4,Zn_2)As_4S_{13}^{*}$	$-\frac{8}{5}$	1	1	$+\frac{3}{5}$	+1
$Cu_6(Ag_4,Zn_2)As_4S_{13}^{*}$	$-\frac{7}{5}$	1	1	$+\frac{2}{5}$	−1

Table 1.
Vertices of Fahlore composition-ordering space (parentheses in formulae indicate tetrahedral metal sites).

Finally, miscibility gaps between high- and low-Ag fahlores in the $(Cu,Ag)_{10}(Fe,Zn)_2(Sb,As)_4S_{13}$ cube (**Figure 7**) at a given temperature (and pressure) are readily calculated using the common tangent method. For the $Cu_{10}Fe_2Sb_4S_{13}$-$Ag_{10}Fe_2Sb_4S_{13}$, $Cu_{10}Zn_2Sb_4S_{13}$-$Ag_{10}Zn_2Sb_4S_{13}$, $Cu_{10}Fe_2As_4S_{13}$-$Ag_{10}Fe_2As_4S_{13}$, and $Cu_{10}Zn_2As_4S_{13}$-$Ag_{10}Zn_2As_4S_{13}$ binary joins common tangents to the Gibbs energy curves are readily found by determining the two values of X_4 that obtain if there is a the crossing of the curve for the chemical potentials of the Cu and Ag endmember components as a function of X_4 at a given temperature and pressure. If this crossing exists these are the two value of X_4 of the fahlores that define the binodal compositions of the miscibility

---

for $X_2 = 0$, $X_3 = 0$, $X_4 = \frac{3}{5}$ and $s = +1$, or

$\mu_j = \overline{G}^o_{(Ag)_6^{TRG}(Cu)_4^{TET}(Fe)_2^{TET}Sb_4S_{13}}$.

And for completeness we may readily add to the RHS of the equation for μ_j the terms we previously omitted that contain the regular solution parameters (W^{TET}_{FeZn}, W^{TRG}_{AgCu}, W^{TET}_{AgCu}, W^{SM}_{AsSb}) and reciprocal parameters ($\Delta\overline{G}^o_{23}$, $\Delta\overline{G}^o_{24}$, $\Delta\overline{G}^o_{34}$):

$$\text{Extra terms on RHS} = W^{TET}_{FeZn}(X_2^2) + W^{TRG}_{AgCu}\left[(1 - 2X_4 + X_4^2) + \frac{1}{25}(20X_4s - 20s + 4s^2)\right]$$
$$+W^{TET}_{AgCu}\left[X_4^2 - \frac{1}{25}(30X_4s - 9s^2)\right] + W^{SM}_{AsSb}(X_3^2) - \Delta\overline{G}^o_{23}(X_2)(X_3)$$
$$+\Delta\overline{G}^o_{24}(X_2)\left(\frac{3}{5} - X_4\right) + \Delta\overline{G}^o_{34}(X_3)\left(\frac{3}{5} - X_4\right).$$

terms which are each equal to zero for $X_2 = 0$, $X_3 = 0$, $X_4 = \frac{3}{5}$ and $s = +1$.

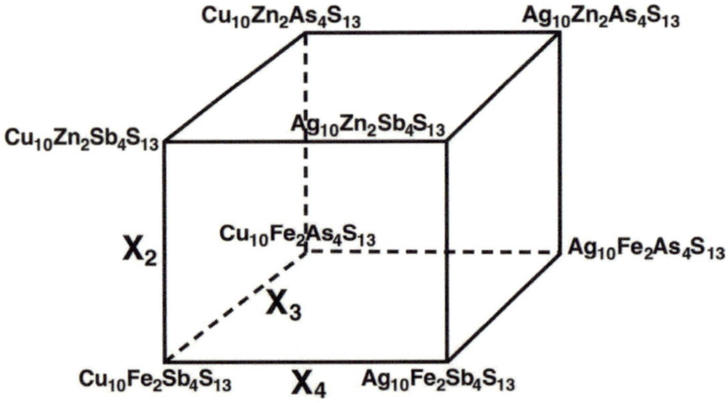

Figure 7.
The $(Cu,Ag)_{10}(Fe,Zn)_2(Sb,As)_4S_{13}$ fahlore cube with X_2 = Zn/(Zn + Fe), X_3 = As/(As+Sb), and X_4 = Ag/(Ag + Cu).

gap at this temperature and pressure. For ternary $(Cu,Ag)_{10}(Fe,Zn)_2Sb_4S_{13}$ and $(Cu,Ag)_{10}(Fe,Zn)_2As_4S_{13}$ fahlores (that is, fahlores on the front and back faces of the fahlore cube in **Figure 7**) binodal pairs may be determined following the same procedure for pseudo-binaries constructed to be at constant $\mu_{Zn(Fe)_{-1}}$ exchange potential (that is for example, in Fe-Zn exchange equilibrium with a (Zn,Fe)S sphalerite with a given Zn/Fe ratio or a given hydrothermal fluid composition). Finally, for quaternary fahlores in the $(Cu,Ag)_{10}(Fe,Zn)_2(Sb,As)_4S_{13}$ cube the identical procedure may be used to determine binodal pairs for pseudo-binaries constructed for constant exchange potentials of $\mu_{Zn(Fe)_{-1}}$ and $\mu_{As(Sb)_{-1}}$ as illustrated by Sack [19] (**Figure 8**).

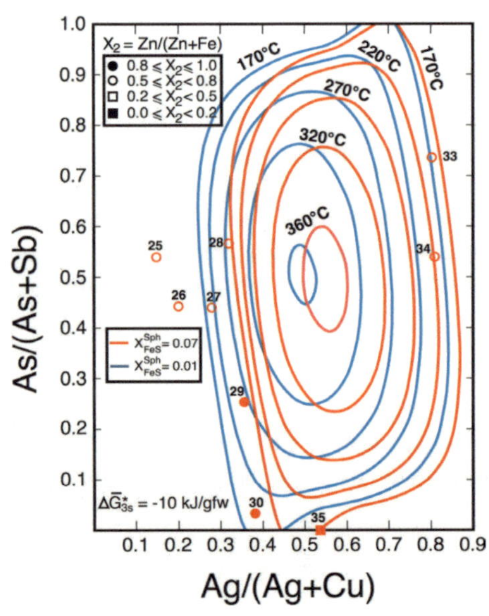

Figure 8.
Calculated 170, 220, 270, 320 and 360°C miscibility gaps for $(Cu,Ag)_{10}(Fe,Zn)_2(Sb,As)_4S_{13}$ fahlores in Fe-Zn exchange equilibrium with (Zn,Fe)S sphalerites with X_{FeS}^{SPH} of 0.01 and 0.07 (from Sack [19]). Red Circles and squares are inclusions of fahlores in chalcopyrites and galenas from the Ag-Pb-Zn ores from the Hope Mine ([108], table I). From Sack [19].

2.3 Gibbs energy of As fahlores

Previous attempts to constrain the Gibbs energy of As-endmember fahlores (for example, Seal et al. [110]) suffer from serious deficiencies. Among these deficiencies were inadequate thermochemical models for activity-composition relations in As-rich fahlores and of those of coexisting sulfides, uncertanties in the thermochemical properties of coexisting phases at the temperatures over which equilibrium may have prevailed in the relevant assemblages, and inadequate accounting for compositional changes in minerals due to retrograde reactions.

The first of these deficiencies has now been addressed. Sack [19] inferred that Ag is stabilized in the trigonal-planar metal sites in As-fahlores relative to Sb-fahlores (that is, $\Delta\overline{G}^*_{3s}$, is negative in the condition of homogeneous equilibrium, Eq. (75)). This inference is required to produce a value of the Gibbs energy of the reciprocal reaction expressing the incompatibility between As and Ag in the fahlore structure, $\Delta\overline{G}^o_{34}$ of reaction (4), positive enough to produce miscibility gaps sufficiently extensive to account for the gaps in Ag/(Ag + Cu) and As/(As+Sb) ratios in As-rich fahlores that occur as inclusions in chalcopyrite and galena grains in the Ag-Pb-Zn ores from the Ag-Pb lode of the Hope Mine on Little Sark in the Channel Islands ([108]; compare **Figure 8**). These parameters are negatively correlated through the tight experimental brackets on the Ag/(Ag + Cu) ratios of $(Cu,Ag)_{10}(Fe,Zn)_2(Sb,As)_4S_{13}$ fahlores in Ag $(Cu)_{-1}$ exchange equilibrium with the assemblage chalcopyrite, pyrite and electrums with various Ag/Au ratios at 300 and 400°C of Refs. [100, 101]. A value of $\Delta\overline{G}^*_{3s}$ of about −10 kJ/gfw results in calculated miscibility gaps that are roughly symmetrically disposed inside the gaps in the compositions of these fahlore inclusions for a temperature of about 250°C and mole fractions of FeS in coexisting sphalerites, X^{SPH}_{FeS}, in the range observed in the Hope Mine ores (0.016–0.07), and a temperature of at least 250° C is required to account for the composition of As-poor fahlore #30 (for example, [18, 23]). This estimate for $\Delta\overline{G}^*_{3s}$ produces a standard error in the estimate of $\Delta\overline{G}^o_{34}$ determined from the experimental constraints of less than 2% of its value, a standard error in $\Delta\overline{G}^o_{34}$ which rises progressively to exceed 5 percent of its value with increase in the assumed value of $\Delta\overline{G}^*_{3s}$ to +10 kJ/gfw ([19], fig. 6). Given these consistencies and previous constraints on the thermochemical model, it may be concluded that the activity-composition relations of $(Cu,Ag)_{10}(Fe,Zn)_2(Sb,As)_4S_{13}$ fahlores have now been established to an excellent approximation (compare **Table 2**).

We may complete the thermochemical model for $(Cu, Ag)_{10}(Fe, Zn)_2(Sb, As)_4S_{13}$ fahlores by specifying the Gibbs energy of formation of one As-endmember, $Cu_{10}Fe_2As_4S_{13}$, $Cu_{10}Zn_2As_4S_{13}$, $Ag_{10}Fe_2As_4S_{13}$, or $Ag_{10}Zn_2As_4S_{13}$, as only one of these is independent of the others in the model (compare [19]). $Cu_{10}Zn_2As_4S_{13}$ is the obvious choice here, as Bryndzia (personal communication to Seal, 1988) reported a preliminary value for the enthalpy of formation of it from the simple sulfides Cu_2S, ZnS, and As_2S_3:

$$\underset{\text{Chalcocite}}{5\ Cu_2S} + \underset{\text{Sphalerite}}{2\ ZnS} + \underset{\text{Orpiment}}{2As_2S_3} \rightleftharpoons \underset{\text{Fahlore}}{Cu_{10}Zn_2As_4S_{13}}, \quad (80)$$

obtained from high temperature synthesis calorimetry. Although the preliminary value obtained was reproducible, − 308.0 ± 1.9 (1σ) kJ/gfw, reaction (80) did not go to completion in the limited times of the calorimetric measurements and minor quantities of additional phases formed, an observation fully consistent with the inference

Parameter	Value (kJ/gfw)
$\Delta\overline{G}^o_{23}$	10.84
$\Delta\overline{G}^o_{24}$	9.00
$\Delta\overline{G}^o_{34}$	61.817
$\Delta\overline{G}^*_s$	−1.6736
$\Delta\overline{G}^*_{2s}$	9.00
$\Delta\overline{G}^*_{3s}$	−10
$\Delta\overline{G}^*_{4s}$	−10.88
W^{TET}_{FeZn}	0.00
W^{SM}_{AsSb}	16.74
W^{TRG}_{AgCu}	0.00
W^{TET}_{AgCu}	29.01

Table 2.
Values of thermodynamic mixing parameters.

that longer reaction times and grinding between annealing cycles are required to produce homogeneous fahlores, react out metastable phases and produce yields of fahlore approaching 100% (for example, [99–102]). Accordingly, this estimate should be regarded with considerable caution, but we might tentatively adopt it and make the additional simplifying assumptions made by Seal et al. [110] that the entropy of reaction (81) is zero at all the temperatures we consider (that is, both the heat capacity of and entropy of $Cu_{10}Zn_2As_4S_{13}$ fahlore are equal to that given by the sum of the simple sulfides in reaction (80)). Seal et al. [110] document that summing the entropies of component sulfides, without adjusting for structural differences, provides more correct estimates for the entropies of chalcopyrite ($CuFeS_2$), proustite (Ag_3SbS_3) and smithite ($AgAsS_2$) than more complex methods proposed by Refs. [111–113]. Sack and Lichtner [18] reached similar conclusions for $(Cu,Ag)_{10}(Fe,Zn)_2Sb_4S_{13}$ fahlore endmembers based on experimental, petrological and geochemical constraints, as their inferred entropies are much closer to those given by the component simple sulfide sums than those which would be achieved, if explicit provisions were made for the configurational entropy in these fahlore endmembers ($\overline{S}^{IC} = -R\,ln\left[\left(\frac{2}{3}\right)^4\left(\frac{1}{3}\right)^2\right]$ = 31.75 J/K-gfw) in computing the appropriate sums.

To impose internal consistency on end member Gibbs energies in the thermodynamic model, we recognize that the entropy of the reciprocal reaction

$$Cu_{10}Fe_2As_4S_{13} + Cu_{10}Zn_2Sb_4S_{13} = Cu_{10}Fe_2Sb_4S_{13} + Cu_{10}Zn_2As_4S_{13}, \tag{81}$$

($\Delta\overline{S}^o_{23}$), and reciprocal reactions (3) and (4) are zero and that the enthalpies of these reciprocal reactions ($\Delta\overline{H}^o_{23}$, $\Delta\overline{H}^o_{24}$ and $\Delta\overline{H}^o_{34}$) require that the enthapies of the remaining As-endmember components are given by the relations

$$\overline{H}^o_{Cu_{10}Fe_2As_4S_{13}} \equiv \overline{H}^o_3 = \overline{H}^o_1 + \overline{H}^o_{Cu_{10}Zn_2As_4S_{13}} - \overline{H}^o_2 - \Delta\overline{H}^o_{23} \tag{82}$$

	$\Delta\overline{H}_f$	$\Delta\overline{S}_f$
	(kJ/gfw)	(J/K-gfw)
$Cu_{10}Fe_2Sb_4S_{13}$[1]	−140.2561	0.565
$Cu_{10}Fe_2Sb_4S_{13}$[2]	−139.461	1.894
$Cu_{10}Zn_2Sb_4S_{13}$	−120.8817	8.309
$Ag_{10}Fe_2Sb_4S_{13}$[1]	−47.6869	0.621
$Ag_{10}Fe_2Sb_4S_{13}$[2]	−46.8919	1.950
$Ag_{10}Zn_2Sb_4S_{13}$	−19.3124	8.365
$Cu_{10}Fe_2As_4S_{13}$[1]	−338.2144	−7.744
$Cu_{10}Fe_2As_4S_{13}$[2]	−337.4193	−6.415
$Cu_{10}Zn_2As_4S_{13}$	−308.0000	0.000
$Ag_{10}Fe_2As_4S_{13}$[1]	−183.8282	−7.688
$Ag_{10}Fe_2As_4S_{13}$[2]	−183.0331	−6.359
$Ag_{10}Zn_2As_4S_{13}$	−144.6138	0.056
FeS[1]	−1.0188	−3.872
FeS[2]	−0.6213	−3.207

Between 177 and 435°C the reference simple sulfides are bcc-Ag_2S, hcp-Cu_2S, ZnS sphalerite, FeS troilite, As_2S_3 orpiment and Sb_2S_3 stibnite.
[1]*T > 324.85°C.*
[2]*324.85–177°C, temperature of bcc-Ag_2S transition to acanthite.*

Table 3.
Enthalpies and entropies of formation of $(Cu,Ag)_{10}(Fe,Zn)_2(Sb, As)_4S_{13}$ fahlore endmembers and FeS sphalerite from simple sulfides at 1 bar and between 177 and 435°C.

$$\overline{H}^o_{Ag_{10}Fe_2As_4S_{13}} = \Delta\overline{H}^o_{34} - \overline{H}^o_1 + \overline{H}^o_3 + \overline{H}^o_4 \tag{83}$$

$$\overline{H}^o_{Ag_{10}Zn_2As_4S_{13}} = \overline{H}^o_{Cu_{10}Zn_2As_4S_{13}} + \overline{H}^o_{Ag_{10}Zn_2Sb_4S_{13}} + \overline{H}^o_{Ag_{10}Fe_2As_4S_{13}} + \overline{H}^o_1 - \overline{H}^o_2 - \overline{H}^o_3 - \overline{H}^o_4 \tag{84}$$

(compare [109]; **Table 3**).

3. Sulfur vacancy (□) bearing fahlores

Recently there has been increasing interest in fahlores exhibiting significant substitution of vacancies for sulfur or $\square(S)_{-1}$. In particular, it has been suggested that the $\square(S)_{-1}$ substitution is the result of the reaction

$$^{TRG}(Ag, Cu)^{+6}_6 + {}^{OCT}S^{-2} = (Ag, Cu)^{+4}_6 + {}^{OCT}\square, \tag{85}$$

a substitution that produces molecular clusters of the metal atoms that octahedrally coordinate the sulfur atoms in sulfur vacancy-free fahlore, and that reduces the charge on the molecular cluster from +6 to +4 to maintain charge balance (for example, [114–117]). This reaction takes place under conditions of low sulfur fugacity and would result in fahlores approximating the formula $(Ag,Cu)_{10}(Fe,Zn)_2(Sb,As)_4S_{12}$ by the reaction

$$(Ag, Cu)_{10}(Fe, Zn)_2(Sb, As)_4S_{13} = (Ag, Cu)_{10}(Fe, Zn)_2(Sb, As)_4S_{12}\square^{OCT} + \tfrac{1}{2}S_2, \quad (86)$$

if only octahedral vacancies for sulfur are produced at low sulfur fugacities. However [26] have determined that sulfur vacancies range to over 2.5 per formula unit in fahlores from the Ag-Pb-Zn Mangazeyskoye ore deposit (Sakha, Russia), requiring that sulfur vacancies are also produced on the 12 tetrahedrally coordinated sulfur sites at low sulfur fugacaties. Accordingly, the reaction describing the reduction of the charge on the molecular clusters must be written as

$$^{TRG}(Ag, Cu)_6^{+6} + n^{VI}S^{-2} + n^{IV}S^{-2} = (Ag, Cu)_6^{(+6-2n\square_{Tot})} + n^{VI}\square + n^{IV}\square \quad (87)$$

where $n\square_{Tot} = n^{VI}\square + n^{IV}\square$. And this reaction would limit $n\square_{Tot}$ to a maximum of 3, as these molecular clusters would then be Ag-Cu alloys with zero charge according to this mechanism. Sack et al. [26] demonstrated that this mechanism is potentially viable by showing that the Mangazeyskoyean fahlores have no net total charge within analytical uncertainty, when the charge on the molecular clusters are calculated by reaction (87). Their analysis thus suggests that a maximum of three sulfur vacancies are permitted in the fahlore structure unless there is another mechanism to account for the $\square(S)_{-1}$ substitution.

This then begs a question about the calibration of the homogeneous reaction that relates the concentration of vacancies on octahedral and tetrahedral sulfur sites in fahlores:

$$(Ag, Cu)_{10}(Fe, Zn)_2(Sb, As)_4S_{12}\square^{OCT} = (Ag, Cu)_{10}(Fe, Zn)_2(Sb, As)_4(S_{11,\square})^{TET}S. \quad (88)$$

Although a rigorous thermodynamic analysis of the equilibrium associated with this reaction is not possible at this point given the severe paucity of relevant data, Sack et al. [26] proved that vacancies strongly prefer the octahedral sulfur site relative to tetrahedral sulfur sites by calculating the upper Ag/(Ag + Cu) displacements of the isotherms of the Ag fahlore breakdown reaction given in **Figure 3**

$$\underset{\text{fahlore}}{Ag_{10}Zn_2Sb_4S_{13}} \rightleftharpoons 3\ \underset{\text{pyrargyrite}}{Ag_3SbS_3} + 2\ \underset{\text{sphalerite}}{ZnS} + \underset{\text{miargyrite (in galena)}}{AgSbS_2} \quad (89)$$

required to account for the Ag/(Ag + Cu) ratios of the Mangazeyskoyean fahlores for the simplifying assumptions that the $\square(S)_{-1}$ substitutions are ideal on the tetrahedral and octahedral sites and that only the configurational effects of these $\square(S)_{-1}$ substitutions on the chemical potential of $Ag_{10}Zn_2Sb_4S_{13}$ need to be considered.

Thus the analysis of Sack et al. [26] provides a qualitative confirmation of the inferences of mineralogists that the $\square(S^{VI})_{-1}$ substitution is by far the dominant vacancy substitution mechanism when $n\square_{Tot} < 1$ pfu, but this community has not yet acknowledged the existence of the $\square(S^{IV})_{-1}$ substitution in fahlore. This is disappointing because it is likely that the extent of operation of this latter substitution probably determines the boundary of the fahlore composition space at low sulfur fugacities. Accordingly, it will be necessary to experimentally determine these boundaries or find evidence for them where fahlores coexist with their breakdown products in ore assemblages. And as emphasized by Sack et al. [26], it appears unlikely that the intrinsic limit of 3 sulfur vacancies per formula unit is reached by natural fahlores as this would result in all of the octahedral metal clusters formed by the

$\square(S^{VI})_{-1}$ substitution becoming (Ag,Cu) alloy nanoparticles completely electrostatically unattached to the fahlore structure. The resulting fahlores would almost certainly have interesting thermoelectric properties with potential applications to waste heat recovery (for example, [118]).

Although the approximation which [26] used to estimate the increases in the Ag/(Ag + Cu) ratios of fahlores coexisting with pyrargyrite, sphalerite and galena/miargyrite with the extent of operation of the $\square(S)_{-1}$ substitution appears adequate to explain the composition data for the Mangazeyskoyean fahlores, a more comprehensive analysis is required to develop a rigorous thermodynamic model for vacancy-bearing fahlores. Developing such a model at present would certainly be a difficult challenge in the light of the paucity of experimental data relating sulfur vacancy contents to sulfur fugacity and the absence of a comprehensive formulation for the thermodynamic properties of these fahlores. Nevertheless, it appears that certain assumptions in the approximation of Sack et al. [26] are inadequate in detail, such as the assumption of ideal mixing of vacancies and sulfur on both octahedral and tetrahedral sites. Although there appears to be no reason to expect that this is not a reasonable assumption in the former case, it does not appear to be so in the latter case, as the Ag/(Ag + Cu) ratios of Mangazeyskoyean fahlores of a given Zn/(Zn + Fe) ratio achieve maximal values when $\square(S)_{-1} \approx 1$ pfu, implying that mixing of vacancies and sulfur on tetrahedral sites is substantially nonideal (that is, $W^{IV}_{\square-S} >> 0$). Then there are the large number of additional parameters that will need to be calibrated to obtain a rigorous thermodynamic model for $(Ag,Cu)_{10}(Fe,Zn)_2(Sb,As)_4S_{(13-x)}$ fahlores with $0 < x < 3$, as well as for fahlores that display stoichiometric deviations from this ideal formula (for example, [119]).

4. Conclusions

As noted in the *Introduction*, metal zoning of fahlores has been well established for the Ag-rich Sb-fahlores in the Ag-Pb-Zn ores from the Keno Hill mining district in Canada [10] and the fissure-vein deposits from the Casalpalca Pb-Zn-Cu-Ag Mine in Peru [8]. Advances in thermochemical models of $(Ag,Cu)_{10}(Zn,Fe)_2(Sb,As)_4S_{13}$ fahlores calibrated based on experimental and petrological studies have led to the determination of the Gibbs energies of the reciprocal reactions expressing the incompatibilities between Zn and Ag and between As and Ag in fahlores, (3) and (4), respectively, and these are the largest contributors to this zoning. In the case of the Keno Hill fahlores, many of which approximate the $(Ag,Cu)_{10}(Zn,Fe)_2Sb_4S_{13}$ subsystem by not having extensive enough sulfur vacancies to significantly alter phase equilibrium (for example, ([26], their **figure 8**)), calibration of reaction equilibria for the fahlore + sphalerite + pyrargyrite + miargyrite assemblage [23] afforded identification of primary high-Ag fahlore compositions, as six out of eight of the individual mine fahlore composition averages determined by Lynch [10] for such fahlores fall simultaneously within the 250–310°C isotherms for $(Ag,Cu)_{10}(Zn,Fe)_2Sb_4S_{13}$ fahlores in this assemblage (compare **Figure 3**), and this is the temperature range determined for primary fluid inclusions in quartz and siderite precipitated during the main stage of ore mineralization [27]. Given this identification of primary high-Ag fahlore compositions, a thermodynamic database for the minerals in the Keno Hill ores, constraints on salinities and calibrations of the thermodynamic properties of hydrothermal fluids, Sack and Lichtner [18] were able to constrain the characteristics of the primary fluids from the eastern and western parts of this 15–20 km fahlore-bearing portion of this more extensive hydrothermal system.

The thermodynamic analysis which led to the construction of **Figure 3** was also useful in identifying fahlores whose compositions reflected significant retrograde evolution, including those from the Husky mine where evidence of unmixing in Ag-fahlores was first observed [21], confirming the earlier prediction of these miscibility gaps based on a thermodynamic analysis of the Fe-Zn exchange equilibrium between fahlore and sphalerite [28]. Although primary fahlore compositions are often not preserved in mesothermal deposits such as those in the Keno Hill District, they may be in epithermal deposits. A clearcut case making this point are the primary fahlores from the Biangiadajuan Ag-Pb-Zn deposit (Inner Mongolia, China) reported by Zhai et al. [24] which were mentioned in the *Introduction*. These have temperatures inferred from the fahlore isotherms in **Figure 3** that correspond to those determined from the earlier fluid inclusion studies of Refs. [34, 35]. Another clearcut demonstation of the viability of the fahlore isotherms in **Figure 3** is provided by the fahlores from the Patricia Zn-Pb-Ag epithemal ore deposit ([31], fig. 14). Nearly all of these fahlores have temperatures between 230 and 140°C according to **Figure 3** with the vast majority of them recording temperatures below 200° C. These temperatures correspond well with those determined from primary fluid inclusions in quartz and sphalerite coexisting with these fahlores, 200 to 140°C, with fahlores having temperatures above 200°C in **Figure 3** possibly reflecting re-equilibration with later higher temperature fluids that characterize the post-ore stage.

The well preserved stratigraphy of crustification bands in the fissure veins in the Casapalca Pb-Zn-Cu-Ag mine record composition paths of $(Cu,Ag)_{10}(Zn,Fe)_2(Sb,As)_4S_{13}$ fahlores which display increasing Ag/(Ag + Cu) and Sb/(Sb + As) with height in the mine with earliest deposited fahlores having higher Sb/(Sb + As) than those deposited at later times (compare **Figure 2**). These trends of concave upward paths on this Ag(Sb) diagram provide ample confirmation of the large Gibbs energy of the reciprocal reaction expressing the incompatibility of As and Ag in fahlore, reaction (4), inferred from the thermochemical analysis of experimental and petrological data summarized here. Operating under the assumption that these fahlores represented their primary compositions motivated [7, 8] to draw envelopes connecting early and late deposited fahlores, and Sack [14] attempted to model the evolution of the fluid-solid system incorporating nonideal properties of fahlore such as a large Gibbs energy of reciprocal reaction (4) into the analysis. However, subsequent petrologic studies of fahlores in the Coeur d'Alene Mining district, Idaho, USA [15, 16] and the Julcani Mining District, Angaraes, Peru [4] demonstrated that the primary fahlore compositions were not preserved, because they were altered by post mineralization retrograde reaction (2) which enriched fahlore in Ag and produced bournonite mineralization as a result of the instability of the $AgSbS_2$ in galena with cooling. Based on their description of the Casapalca deposit of [7, 8], it appears that retrograde reactions also modified fahlore compositions there, as some fahlores are enriched in Fe as a consequence of the breakdown of chalcopyrite by a reaction such as that suggested by Wu and Petersen [8] (compare reaction (1)) and bournonite is a late stage mineral, suggesting that primary fahlore was enriched in Ag by reaction (2), as there is abundant galena in these samples and no Ag in galena was reported.

For the purpose of reconstructing the primary compositions of fahlore and galena from galena-rich ores from the Coeur d'Alene Mining District, [15] chose a galena-rich sample from the West Chance vein of the Sunshire Mine (WC-7) and one from the

Gold Hunter vein from the Lucky Friday Mine (GH-5) and used digitized back scattered electron (BSE) images to determine the modal abundance of galena, fahlore and bournonite in sample WC-7, and of galena, fahlore, bournonite and sphalerite in sample GH-5. From the molar proportions of galena, fahlore and bournonite and the assumption that all of the bournonite in these sample was produced by reaction (2) they determined that the molar Ag/Ag + Cu) ratios of the primary fahlores in these samples were 0.1100 (WC-7) and 0.1022 (GH-5), as opposed to their present values of 0.2767 and 0.2860, respectively. Using these estimates, those for the primary $AgSbS_2$ contents of galena, the characteristics of the rest of the mineral assemblage, and constraints on temperature and fluid salinity, Sack and Lichtner [18] constrained the compositional characteristics of the high temperature hydrothermal fluids which equilibrated with these samples. And to further prove that reaction (2) was responsible for producing the most Ag-rich fahlores found in the Coeur d'Alene Mining District rather than the hydrothermal fractionation proposed by Refs. [9, 120], Sack et al. [16] examined one of the samples with the most Ag-rich fahlores examined by Hackbarth [120] from the Galena Mine from the collection of her PhD samples at the Harvard Museum, as well as diamond drill cores from the mine containing contacts between fahlore-siderite veins and cross-cutting galena shoots obtained from the mine operator Coeur Silver Valley Inc. From these studies they determined that the sample from the Galena mine with fahlore with the highest Ag/(Ag + Cu) ratios examined by Hackbarth [120] contained abundant bournonite and that fahlores at the contacts between the siderite-fahlore veins and the cross-cutting galena shoots in the drill cores had elevated Ag/(Ag + Cu) ratios relative to those in siderite-fahlore veins not in contact with galena shoots.

It should be clear from the above discussion that comprehensive modeling of the metal zoning produced at given times and temperature gradients will depend not only on being able to constrain original fluid compositions from reconstructed mineral compositions and assemblages, but also on being able to effectively model reactive transport (for example, [36]). Refs. [18, 22] outlined an approach to constrain original hydrothermal fluid compositions utilizing the solid state thermodynamic database touched on here and the models for the thermodynamic properties of hydrothermal fluids such as those in programs and databases in SUPCRT92 [121] and SOLTHERM [122]. Computer codes to model reactive transport are also available (for example, PFLOTRAN). Even in the absence of simulations combining all of these elements, the basic features of metal zoning in fahlores may be anticipated given knowledge that the Gibbs energies of the reciprocal reactions expressing the incompatibilities of Zn and As, Zn and Ag, and As and Ag in fahlore, $\Delta\overline{G}^{o}_{23}$, $\Delta\overline{G}^{o}_{24}$ and $\Delta\overline{G}^{o}_{34}$, are large. Because these reciprocal energies are approximately temperature independent, sector zoned fahlores may also provide useful information about depositional conditions on their own, if their primary sector compositions are preserved (for example, [98, 123]). Finally, an accurate value for the Gibbs energy of formation from the simple sulfides of at least one As-fahlore endmember must be obtained to complete the thermodynamic model for $(Ag,Cu)_{10}(Zn,Fe)_2(Sb,As)_4S_{13}$ fahlores and its formulation must be extended to account for sulfur vacancy-bearing fahlores approximating the formula $(Cu,Ag)_{10}(Fe,Zn)_2(Sb,As)_4S_{(13-x)}$ with $0 < x < 3$ for calibration when additional experimental and natural data become available to constrain the limits of the fahlore stability field at low sulfur fugacities. Characterizing the material properties of fahlores at these limits should be quite interesting and productive.

Acknowledgements

I gratefully acknowledge discussions with Lisa Hardy and Peter Lichtner, the comments of Phil Goodell on an earlier draft, the suggestions of the editor Miloš René and the able assistance of Publishing Process Manager Iva Horvat.

I also acknowledge the support of the OFM staff Odester, Fifo, and Margie.

Author details

Richard Sack
OFM Research Southeast, Thompson's Station, Tennessee, USA

*Address all correspondence to: fahlore@gmail.com

References

[1] Stanton RL. Ore Petrology. New York: McGraw-Hill; 1972. 713 p. ISBN: 0070608431

[2] Goodell PC. Zoning and paragenesis in the Julcani district, Peru: A study of metal ratios [PhD thesis]. Cambridge: Harvard University; 1970

[3] Goodell PC, Petersen U. Julcani mining district, Peru: A study of metal ratios. Economic Geology. 1974;**69**: 347-361. DOI: 10.2113/gsecongeo.69.3.347

[4] Sack RO, Goodell PC. Retrograde reactions involving galena and Ag–sulfosalts in a zoned ore deposit, Julcani, Peru. Mineralogical Magazine. 2002;**66**: 1043-1062. DOI: 10.1180/0026461026660076

[5] Barton PB Jr, Toulmin P III. Phase relations involving sphalerite in the Fe–Zn–S system. Economic Geology. 1966; **61**:815-849. DOI: 10.2113/gsecongeo.61.5.815

[6] Barton PB Jr, Bethke PM, Roedder E. Environment of ore deposition in the Creede mining district, San Juan Mountains, Colorado: III. Progress toward interpretation of the chemistry of the ore-forming fluid for the OH vein. Economic Geology. 1977;**72**:1-24. DOI: 10.2113/gsecongeo.72.1.1

[7] Wu I. Geochemistry of tetrahedrite-tennantite at Casapulca, Peru [PhD thesis]. Cambridge: Harvard University; 1975

[8] Wu I, Petersen U. Geochemistry of tetrahedrite-tennantite at Casapalca, Peru. Economic Geology. 1977, 1977;**72**: 993-1016. DOI: 10.2113/gsecongeo.72.6.993

[9] Hackbarth CJ, Petersen U. Systematic compositional variations in argentian tetrahedrite. Economic Geology. 1984; **79**:448-460. DOI: 10.2113/gsecongeo.79.3.448

[10] Lynch JVG. Large-scale hydrothermal zoning reflected in the tetrahedrite-freibergite solid solution, Keno Hill Ag-Pb-Zn district, Yukon. Canadian Mineralogist. 1989a;**27**: 383-400

[11] McKinistry HE, Noble JA. The veins of Casapalca, Peru. Economic Geology. 1932;**27**:501-522. DOI: 10.2113/gsecongeo.27.6.501

[12] Rye RO, Ohmoto H. Sulfur and carbon isotopes and ore genesis: A review. Economic Geology. 1974;**69**: 826-842. DOI: 10.2113/gsecongeo.69.6.826

[13] Rye RO, Sawkins FJ. Fluid inclusion and stable isotope studies on the Casapalca Ag-Pb-Zn-Cu deposit, Central Andes, Peru. Economic Geology. 1974; **69**:181-205. DOI: 10.2113/gsecongeo.69.2.181

[14] Sack RO. Thermochemistry of tetrahedrite-tennantite fahlores. In: Price GD, Ross NL, editors. The Stability of Minerals. London: Chapman and Hall; 1992. pp. 243-266. ISBN: 0-412-44150-0

[15] Sack RO, Kuehner SM, Hardy LS. Retrograde Ag-enrichment in fahlores from the Coeur d'Alene mining district, Idaho, USA. Mineralogical Magazine. 2002;**66**:215-229. DOI: 10.1180/0026461026610024

[16] Sack RO, Fredericks RM, Hardy LS, Ebel DS. Origin of high-silver fahlores from the Galena Mine, Wallace, Idaho, USA. American Mineralogist. 2005;**90**: 1000-1007. DOI: 10.2138/am.2005.1651

[17] Knowles CR. A microprobe study of silver ore in northern Idaho. In: Gooley R, editor. Microbeam Analysis. San Fransisco: San Fransisco Press; 1983. pp. 61-64

[18] Sack RO, Lichtner PC. 2009, Constraining compositions of hydrothermal fluids in equilibrium with polymetallic ore-forming sulfide assemblages. Economic Geology. 2009; **104**:1249-1264. DOI: 10.2113/gsecongeo.104.8.1249

[19] Sack RO. Fahlore thermochemistry: Gaps inside the $(Cu,Ag)_{10}(Fe,Zn)_2(Sb,As)_4S_{13}$ cube. Petrology. 2017;**25**:498-515. DOI: 10.1134/S0869591117050071

[20] Boyle RW. Keno Hill-Galena Hill lead-zinc-silver deposits. Geological Survey of Canada Bulletin. 1965;**111**. 302 p

[21] Sack RO, Lynch JGV, Foit FF Jr. Fahlore as a petrogenetic indicator: Keno Hill Ag-Pb-Zn district, Yukon, Canada. Mineralogical Magazine. 2003;**67**: 1023-1038. DOI: 10.1180/0026461036750141

[22] Sack RO, Lichtner PC. Erratum. Economic Geology. 2010;**105**:249. DOI: 10.2113/gsecongeo.105.1.249

[23] Sack RO. Internally consistent database for sulfides and sulfosalts in the system Ag_2S–Cu_2S–ZnS–FeS–Sb_2 S_3–As_2S_3. Geochimica et Cosmochimica Acta. 2005;**69**:1157-1164. DOI: 10. 1016/j.gca.2004.08.017

[24] Zhai D, Jiajin L, Cook NJ, Wang X, Yang Y, Zhang A, et al. Mineralogical, textural, sulfur and lead isotope constraints on the origin of Ag-Pb-Zn mineralization at Bianjiadayuan, Inner Mongolia, NE China. Mineralium Deposita. 2019;**54**:47-66. DOI: 10.0007/s00126-018-0804-6

[25] Ebel DS, Sack RO. Experimental determination of the free energies of formation of freibergite fahlores. Geochimica et Cosmochimica Acta. 1994;**58**:1237-1242. DOI: 10.1016/0016-7037(94) 90377-8

[26] Sack RO, Lyubimtseva NG, Bortnikov NS, Anikina EY, Borisovsky SE. Sulfur vacancies in fahlores from the Ag-Pb-Zn Mangazeyskoye Ore Deposit (Sakha, Russia). Contributions to Mineralogy and Petrology. 2022;**177**(8): 1-26. DOI: 10.1007/s00410-022-01942-5

[27] Lynch JVG. Hydrothermal zoning in the Keno Hill Ag–Pb–Zn vein system: a study in structural geology, mineralogy, fluid inclusions and stable isotope geochemistry [Ph.D. thesis]. Alberta: University of Alberta; 1989. DOI: 10.7939/R38K75524

[28] O'Leary MJ, Sack RO. Fe-Zn exchange reaction between tetrahedrite and sphalerite in natural environments. Contributions to Mineralogy and Petrology. 1987;**96**:415-425. DOI: 10.1007/BF01166687

[29] Keighin CW, Honea RM. The system Ag–Sb–S from 600°C to 200°C. Mineralium Deposita. 1969;**4**:153-171. DOI: 10.1007/BF00208050

[30] Makovicky E. Crystal structures of sulfides and other chalcogenides. In: Vaughan DJ, editor. Sulfide Mineralogy and Geochemistry, Mineralogical Society of America, Reviews in Mineralogy and Geochemistry. Vol. 61. Chantilly, Virginia: Mineralogical Society of America; 2006. pp. 7-125. DOI: 10.1515/9781501509490-003

[31] Chinchilla D, Ortega L, Piña R, Merinero R, Moncada D, Bodnar RJ, et al. The Patricia Zn-Pb-Ag epithermal

ore deposit: An uncommon type of mineralization in northeastern Chile. Ore Geology Reviews. 2016;**73**:104-126. DOI: 10.1016/j.oregeorev.2015.10.026

[32] Alves FEA, AVC N, Soares MB, Neuman R, da Silva GM, Varca AC, et al. Genetic implications from textures, mineralogy, and geochemistry: the case of Zona Basal - a singular polymetallic occurrence in the Quadrilatero Ferrifero, Brazil. Contribution to Mineralogy and Petrology. 2022;**177**:1-48. DOI: 10.1007/s00410-022-01913-w

[33] Zhai D. Fluid-rock interactions leading to the formation of the epithermal Ag-Pb-Zn veins: A perspective of thermodynamic modeling. Fundamental Research. 2023; **3**(4):570-578. DOI: 10.1016/j.fmre.2022.03.004

[34] Wang X. The study of the Metallogenic characteristics and genesis of the Bianjiadayuan Ag-Pb-Zn deposit in Inner Mongolia, China [MS thesis]. Beijing: China University of Geosciences; 2014 (in Chinese with English)

[35] Ruan B, Lv X, Yang W, Liu S, Yu Y, Wu C, et al. Geology, geochemistry and fluid inclusions of the Bianjiadayuan Pb-Zn-Ag deposit, Inner Mongolia, NE China: Implications for tectonic setting and metallogeny. Ore Geology Reviews. 2015;**71**:121-137. DOI: 10.1016/j.oregeorev.2015.05.004

[36] Lichtner PC. Continuum model for simultaneous chemical reactions and mass transport in hydrothermal systems. Geochimica et Cosmochimica Acta. 1985; **49**:779-800. DOI: 10.1016/0016-7037(85)90172-3

[37] Margules M. Über die Zusammensetzung der gesätigten dämpfe von mischungen: Sitzungber. Akademie der Wissenschaften in Wien. 1895;**104**:1243-1278

[38] Hildebrand JH. 1929, Solubility, XII, regular solutions. Journal of the American Chemical Society. 1929;**51**: 77-80. DOI: 10.1021/ja01376a009

[39] Scatchard G, Hamer W. 1935 The application of equations for the chemical potentials to partially miscible solutions. Journal of the American Chemical Society. 1935;**57**:1805-1809. DOI: 10.1021/ja01313a016

[40] Carlson HC, Colburn AP. Vapor-liquid equilibria of non-ideal solutions. Industrial and Engineering Chemistry. 1942;**84**:581-589. DOI: 10.1021/ie50389a013

[41] Hardy HK. A "sub-regular" solution model and its applications to some binary alloy systems. Acta Metallurgica. 1950;**1**:202-209. DOI: 10.1016/0001-6160(53)90059-5

[42] Thompson JB Jr. Thermodynamic properties of simple solutions. In: Abelson PH, editor. Researches in Geochemistry. Vol. 2. New York: John Wiley; 1967. pp. 340-361

[43] Lawson AW. On simple binary solutions. Journal of Chemical Physics. 1947;**15**:831-847. DOI: 10.1063/1.1746346

[44] Thompson JB Jr. Chemical reactions in crystals. American Mineralogist. 1969; **54**:341-375

[45] Thompson JB Jr. Chemical reactions in crystals: corrections and clarification. American Mineralogist. 1970;**55**:528-532

[46] Sack RO. Some constraints of the thermodynamic mixing properties of Fe-Mg orthopyroxenes and olivines. Contributions to Mineralogy and

Petrology. 1980;**71**:257-269. DOI: 10.1007/BF00371667

[47] Sack RO, Ghiorso MS. Importance of considerations of mixing properties in establishing an internally consistent thermodynamic database: Thermochemistry of minerals in the system $Mg_2SiO_4 - Fe_2SiO_4 - SiO_2$. Contributions to Mineralogy and Petrology. 1989;**102**:41-68. DOI: 10.1007/BF01160190

[48] Balabin AI, Sack RO. Thermodynamics of (Zn,Fe)S sphalerite: A CVM approach with large basis clusters. Mineralogical Magazine. 2000; **64**(5):923-943. DOI: 10.1180/002646100549751

[49] Sack RO, Ebel DS. Thermochemistry of sulfide mineral solutions. In: Vaughan DJ, editor. Sulfide Mineralogy and Geochemistry. Vol. 61. Reviews in Mineralogy and Geochemistry; 2006. pp. 265-364. DOI: 10.1515/9781501509490-007

[50] Darken LS. Application of Gibbs-Duhem relation to ternary and multicomponent systems. Journal of the American Chemical Society. 1950;**72**: 2909-2914. DOI: 10.1021/ja01163a030

[51] Pauling L, Newmann EW. The crystal structure of binnite (Cu, Fe)$_{12}$As$_4$S$_{13}$ and the chemical composition of minerals of the tetrahedrite group. Zeitschrift für Kristallographie. 1934;**88**:54-62. DOI: 10.1524/zkri.1934.88.1.54

[52] Harlov DE, Sack RO. Thermodynamics of polybasite-pearceite solutions. Geochimica et Cosmochimica Acta. 1994;**58**:4363-4375. DOI: 10.1016/0016-7037(95)00308-M

[53] Ghiorso MS, Evans BW. Thermodynamics of the amphiboles: Ca-Mg-Fe^{2+} quadrilateral. American Mineralogist. 2002;**87**:79-98. DOI: 10.2138/am-2002-0109

[54] Sack RO. Spinels as petrogenetic indicators: activity-composition relations at low pressures. Contributions to Mineralogy and Petrology. 1982;**79**: 169-182. DOI: 10.1007/ BF01132886

[55] Ghiorso MS. Application of the Darken equation to mineral solid solutions with variable degrees of order–disorder. American Mineralogist. 1990; **75**:539-543

[56] Scott SD, Barnes HL. Sphalerite geothermometry and geobarometry. Economic Geology. 1971;**1971**(66): 653-669. DOI: 10.2113/gsecongeo.66.4.653

[57] Hutcheon I. Calculation of metamorphic pressure using the sphalerite-pyrrhotite-pyrite equilibrium. American Mineralogist. 1978;**63**:87-95

[58] Hutcheon I. Calculated phase relations for sphalerite-pyrrhotite-pyrite: Correction. American Mineralogist. 1980;**65**:1063-1064

[59] Hutchison MN, Scott SD. Sphalerite geobarometry in the Cu-Fe-Zn-S system. Economic Geology. 1981;**76**:143-153. DOI: 10.2113/gsecongeo.76.1.143

[60] Hutchison MN, Scott SD. Experimental calibration of the sphalerite cosmobarometer. Geochimica et Cosmochimica Acta. 1983;**47**:101-108. DOI: 10.1016/0016-7037(83)90094-7

[61] Balabin AI, Urusov VS. Recalibration of the sphalerite cosmobarometer: Experimental and theoretical treatment. Geochimica et Cosmochimica Acta. 1995; **59**:1401-1410. DOI: 10.1016/0016-7037(95)00052-2

[62] Martin JD, Gil AS. An integrated thermodynamic mixing model for sphalerite geothermometry from 300 to 850°C and up to 1 Gpa. Geochimica et Cosmochimica Acta. 2005;**69**:995-1006. DOI: 10.1016/j.gca.2004.08.009

[63] Toulmin P III, Barton PB Jr, Wiggins LB. Commentary on the sphalerite geobarometer. American Mineralogist. 1991;**76**:1038-1051

[64] Pankratz LB, King EG. High-temperature heat contents and entropies of two zinc sulfides and four solid solutions of zinc and iron sulfides. Bureau of Mines Report of Investigations. 1965;**6708**:1-8

[65] Spalek J, Lewicki A, Tarnawski Z, Furdyna JK, Galazka RR, Obuszko Z. Magnetic susceptability of semimagnetic semiconductors in the high temperature regime and the role of superexchange. Physical Review B. 1986;**33**:3407-3418

[66] Furdyna JK. Diluted magnetic semiconductors. Journal of Applied Physics. 1988;**64**:R29-R64. DOI: 10.1063/1.3 41700

[67] Twardowski A. Magnetic properties of Fe-based diluted magnetic semiconductors. Journal of Applied Physics. 1990;**67**:5108-5113. DOI: 10.1063/1. 344685

[68] Di Benedetto F, Bernardini GP, Caneschi A, Cipriani C, Danti C, Pardi L, et al. EPR and magnetic investigations on sulphides and sulphosalts. European Journal of Mineralogy. 2002;**14**: 1053-1060. DOI: 10.1127/0935-1221/ 2002/0014-1053

[69] Di Benedetto F, Andreozzi GB, Bernardini GP, Borgheresi M, Caneschi A, Cipriani C, et al. Short range order of Fe^{2+} by ^{57}Mossbauer spectroscopy and magnetic susceptibility. Physics and Chemistry of Minerals. 2005a;**32**:339-348. DOI: 10.1007/s00269-005-0002-9

[70] Kikuchi RD. A theory of cooperative phenomena. Physical Reviews. 1951;**81**: 988-1003. DOI: 10.1103/ PhysRev.81.988

[71] Hijmans J, de Boir J. An approximation method for order-disorder problems I. Physica. 1955;**21**: 471-484. DOI: 10.1016/S0031-8914(54) 92484-2

[72] van Baal CM. Order-disordrer transformations in a generalized Ising alloy. Physica. 1973;**64**:571-586. DOI: 10.1016/0031-8914(73)90010-4

[73] Sanchez JM, de Fontaine D. Ordering in fcc lattices with first- and second-neighbor interactions. Physical Reviews B. 1980;**21**:216-228. DOI: 10.1103/PhysRevB.21.216

[74] Sanchez JM, Ducastelle F, Gratias D. Generalized cluster description of multicomponent systems. Physica A. 1984;**128**:334-350. DOI: 10.1016/ 0378-4371(84)90096-7

[75] de Fontaine D. Cluster approach in order-disorder transformations in alloys. Solid State Physics. 1994;**47**:33-176. DOI: 10.1016/S0081-1947(08)60639-6

[76] Sanchez JM, Mohri T. Approximate solutions to the cluster variation free energies by the variable basis cluster expansion. Computational Materials Science. 2016;**122**:301-306. DOI: 10.1016/j.commatsci.2016.05.035

[77] Fleet ME. Thermodynamic properties of (Zn,Fe)S solid solutions at 850°C. American Mineralogist. 1975;**60**: 466-470

[78] Boorman RS. Subsolidus studies in the ZnS-FeS-FeS_2 system. Economic

Geology. 1967;**62**:614-631. DOI: 10.2113/gsecongeo.62.5.614

[79] Scott SD, Kissin SA. Sphalerite composition in the system Zn-Fe-S system below 300°C. Economic Geology. 1973;**68**:475-479. DOI: 10.2113/gsecongeo.68.4.475

[80] Hollenbaugh DW, Carlson EH. The occurrence of wurtzite polytypes in eastern Ohio. Canadian Mineralogist. 1983;**21**:697-703

[81] Chao GY, Gault RA. The Occurrence of Two Rare Polytypes of Wurtzite, 4H and 8H, at Mont Saint-Hilaire. Vol. 36. Quebec: Canadian Mineralogist; 1998. pp. 775-778

[82] L'Heureux I. Origin of banded patterns in natural sphalerite. Physical Review E. 2000;**62**:3234-3245. DOI: 10.1103/PhysRevE.62.3234

[83] Di Benedetto F, Bernardini GP, Costagliola P, Plant D, Vaughan DJ. Compositional zoning in sphalerite crystals. American Mineralogist. 2005b; **90**:1384-1392. DOI: 10.2138/am.2005.1754

[84] Osachii EG, Gorbaty YE. Raman spectra and unit cell parameters of sphalerite solid solutions ($Fe_xZn_{1-x}S$). Geochimica et Cosmochimica Acta. 2010;**74**:568-573. DOI: 10.1016/j.gca.2009.10.022

[85] Twardowski A, Lewicki A, Arciszewska M, de Jonge WJM, Swagsten HJM, Demianiuk M. Magnetic susceptibility of iron based semimagnetic semiconductors: High temperature regime. Physical Review B. 1988;**38**(15):10749-10754. DOI: 10.1103/PhysRevB.38.10749

[86] Twardowski A, Swagsten HJM, de Jonge WJM, Demianiuk M. Magnetic properties of the diluted magnetic semiconductor $Zn_{1-x}Fe_xS$. Physical Review B. 1991;**44**(5):2220-2226. DOI: 10.1103/PhysRevB.44.2220

[87] Twardowski A, Fries T, Shapira T, Demianiuk M. The d–d exchange interaction in diluted magnetic semiconductors FeZnS. Journal of Applied Physics. 1993;**73**(10):6087-6089. DOI: 10.1063/1.353478

[88] Wright K, Gale JD. A first principles study of the distribution of iron in sphalerite. Geochimica et Cosmochimica Acta. 2010;**74**:3514-3520. DOI: 10.1016/j.gca.2010.03.014

[89] Pring A, Tarantino SC, Tenailleau C, Etschmann B, Carpenter MA, Zhang M, et al. 2008, The crystal chemistry of Fe-bearing sphalerites: An infrared spectroscopic study. American Mineralogist. 2008;**93**:591-597. DOI: 10.2138/am.2008.2610

[90] Guggenheim EA. Thermodynamics: An Advanced Treatment for Chemists and Physicists. Amsterdam, Netherlands: North-Holland Physics Publishing; 1967. 390 p

[91] Chareev DA, Osadchii VO, Shiryaev AA, Nekrasov AN, Koshelev AV, Osadchii EG. Single-crystal Fe-bearing sphalerite: synthesis, lattice parameter, thermal expansion coefficient and microhardness. Physics and Chemistry of Minerals. 2017;**44**: 287-296. DOI: 10.1007/s00269-016-0856-z

[92] de Médicis R. Cubic FeS, a metastable iron sulfide. Science. 1970; **170**:1191-1192. DOI: 10.1126/science.170.3963.1191

[93] Takeno S, Zoka H, Niiharo T. Metastable cubic iron sulfide, with

special reference to mackinawite. American Mineralogist. 1970;**55**: 1639-1649

[94] Lepetit P, Bente K, Doering T, Luckhaus S. Crystal chemistry of Fe-containing sphalerites. Physics and Chemistry of Minerals. 2003;**30**:185-191. DOI: 10.1007/s00269-003-0306-6

[95] Hume-Rothery W. Electrons, Atoms, Metals and Alloys. New York: Dover; 1963. 387 p

[96] Johnson ML, Jeanloz R. A brillouin-zone model for compositional variation in tetrahedrite. American Mineralogist. 1983;**68**:220-226

[97] Jeanloz R, Johnson ML. A note on the bonding, optical spectrum and composition of tetrahedrite. Physics and Chemistry of Minerals. 1984;**11**:52-54. DOI: 10.1007/BF00309375

[98] Raabe KC, Sack RO. Growth zoning in tetrahedrite-tennantite from the Hock Hocking Mine, Alma, Colorado. Canadian Mineralogist. 1984;**22**(4): 577-582

[99] Sack RO, Loucks RR. Thermodynamic properties of tetrahedrite-tennantites: I. Constraints on the interdependencies of the Ag⇔Cu, Fe⇔Zn, Cu⇔Fe, and As⇔Sb exchange reactions. American Mineralogist. 1985; **70**:1270-1289

[100] Ebel DS, Sack RO. Ag–Cu and As–Sb exchange energies in tetrahedrite-tennantite fahlores. Geochimica et Cosmochimica Acta. 1989;**53**:2301-2309. DOI: 10.1016/0016-7037(89)90352-9

[101] Ebel DS, Sack RO. As–Ag incompatibility in fahlores. Mineralogical Magazine. 1991;**55**: 521-528. DOI: 10.1180/minmag.1991.055.381.04

[102] Sack RO, Ebel DS. As–Sb exchange energies in tetrahedrite-tennantite fahlores and bournonite-seligmannite solid solutions. Mineralogical Magazine. 1993;**57**:635-642. DOI: 10.1180/minmag.1993.057.389.07

[103] Sack RO. Note on "large-scale hydrothermal zoning reflected in the tetrahedrite-freibergite solid solution Ag-Pb-Zn District, Yukon" by J.V. Gregory Lynch. Canadian Mineralogist. 2002;**40**:1717-1719. DOI: 10.2113/gscanmin.40.6.1717

[104] Chutas NI, Sack RO. Ore genesis at La Colorada Ag-Zn-Pb deposit in Zacatecas, Mexico. Mineralogical Magazine. 2004;**68**(6):923-937. DOI: 10.1180/0026461046860231

[105] Sack RO, Brackebusch FW. Fahlore as an indicator of mineralization temperature and gold fineness. CIM Bulletin. 2004;**97**:78-83

[106] Hernández ANG, Akasaka M. Silver-bearing and associated minerals in El Zancudo Deposit, Antioquia, Colombia. Resource Geology. 2007; **57**(4):386-399. DOI: 10.1111/j.1751-3928.2007.00030.x

[107] Hernández ANG, Akasaka M. Ag-rich tetrahedrite in the El Zancudo deposit, Colombia: Occurrence, chemical compositions and genetic temperatures. Resource Geology. 2010;**60**(3):218-233. DOI: 10.1111/j.1751-3928.2010.00133.x

[108] Ixer RA, Stanley CJ. Silver mineralization at Sark's Hope Mine, Sark, Channel Islands. Mineralogical Magazine. 1983;**47**:539-545. DOI: 10.1180/minmag.1983.047.345.15

[109] Sack RO, Ebel DS, O'Leary MJ. Tennahedrite thermochemistry and metal zoning. In: Helgeson HC, editor. Chemical Transport in Metasomatic

Processes. Dordrecht: D. Reidel; 1987. pp. 701-731

[110] Seal RR II, Essene EJ, Kelly WC. Tetrahedrite and tennantite: Evaluation of thermodynamic data and phase equilibria. Canadian Mineralogist. 1990; **28**:725-738

[111] Lattimer WM. The Oxidation States of the Elements and their Potentials In Aqueous Solutions. 2nd ed. New York: Prentice-Hall; 1952. 374 p. ASIN: B0006AT5R8

[112] Fyfe WS, Turner FJ, Verhoogen J. Metamorphic Reactions and Metamorphic Facies, Memoir. Vol. 73. Geological Society of America; 1958. 259 p

[113] Grønvold F, Westrum EF Jr. Heat capacities and thermodynamic functions of iron disulfide (pyrite), iron diseienide, and nickel diseienide from 5 to 350 K. The estimation of standard entropies of transition metal chalcogenides. Inorganic Chemistry. 1962;**1**:36-48. DOI: 10.1021/ic50001a009

[114] Rozhdestvenskaya IV, Zayakina NV, Samusikov VP. Crystal structure features from a series of tetrahedrire-freibergite. Mineralgiceskij Zhurnal. 1993;**15**:9-17. (In Russian)

[115] Welch MD, Stanley CJ, Spratt J, Mills SJ. Rozhdestvenskayaite $Ag_{10}Zn_2Sb_4S_{13}$ and argentotetrahedrite $Ag_6Cu_4(Fe^{2+},Zn)_2Sb_4S_{13}$: Two Ag-dominant members of the tetrahedrite group. European Journal of Mineralogy. 2018;**30**(6):1163-1171. DOI: 10.1127/ejm/2018/0030-2773

[116] Biagioni C, George L, Cook N, Makovicky E, Moelo Y, Pasero M, et al. The Tetrahedrite group: Nomenclature and classification. American Mineralogist. 2020;**103**(1):109-122. DOI: 10.2138/am-2020-7128

[117] Qu K, Sun W, Nestola F, Gu X, Yang Z, Sima X, et al. Kenorozhdestvenskayaite-(Fe), $Ag_6(Ag_4Fe_2)Sb_4S_{12}W$: A new tetrahedrite group mineral containing a natural $[Ag_6]^{4+}$ cluster and its relationship to the synthetic ternary phosphide $(Ag_6M_4P_{12})M_6$. American Mineralogist. 2024;**109**:1275-1283

[118] Fan X, Case ED, Lu,X, Morelli DT. Room temperature mechanical properties of natural-mineral-based thermoelectric. Journal of Material Science. 2013; 48(21):7540-7550. DOI: 10.1007/ s10853-013-7569-1

[119] Lyubimtseva NG, Bortnikov NS, Borisovsky SE, Prokofiev VY, Vikent'eva OV. Fahlore and sphalerite from the Darasun gold deposit in the Eastern Transbaikal Region, Russia; I. Mineral assemblages and intergrowths, chemical composition, and its evolution. Geology of Ore Deposits. 2018;**60**(2): 93-120. DOI: 10.1134/ S107570151 8020034

[120] Hackbarth CJ. Depositional modeling of tetrahedrite in the Coeur d'Alene District [PhD thesis]. Cambridge: Harvard University; 1984

[121] Johnson JW, Oelkers H, Helgeson HC. SUPCRT92: A software package for calculating the standard molal thermodynamic properties of minerals, gases, aqueous species, and reactions from 1 to 5000 bar and 0 to 1000°C. Computers and Geosciences. 1992;**18**:899-947. DOI: 10.1016/ 0098-3004(92)90029-Q

[122] Reed MH, Palandri J. SOLTHERM: Data Base of Equilibrium Constants for Aqueous-Mineral-Gas Equilibria. Eugene, OR 97403: Department of

Geological Sciences, University of Oregon; 2023. (Unpublished report)

[123] Lyubimtseva NG, Sack RO, Bortnikov NS, Borisovsky SE, Prokofiev VY, Balashov FV. The zonal fahlore and sphalerite from the Darasun gold deposit, Transbaikalia, Russia: An example of a self-organizing system and their depositional conditions. Geology of Ore Deposits. 2023;**65**(4):346-380. DOI: 10.1134/S1075701523040037

Chapter 5

Alteration of Durbachites and U-Th Mineralization in the Durbachites of the Třebíč Pluton, Czech Republic

Miloš René

Abstract

The Třebíč pluton comprises ultrapotassic (ultra-K) biotite-amphibole plutonic rocks of the durbachite suite. The Třebíč pluton intruded in the eastern Moldanubian Zone of the Bohemian Massif. These igneous rocks remain highly magnesian igneous rocks, markedly enriched in Th, U and K. The described U-Th mineralization is bounded on brittle shear zone evolved in the north-eastern part of the Třebíč pluton near small village Tasov. The ore mineralization in this ore deposit is coupled with intense low-temperature, hydrothermal alterations of these melasyenites (hematitization, chloritization, albitization and carbonatization). These alterations are accompanied by enrichment in Ti, Fe, Ca, and depletion in Si, Ba and Sr. The ore mineralization is formed predominantly of the thorite and coffinite. The rare earth elements (REE) are concentrated in the REE-fluorocarbonate synchysite-(Ce).

Keywords: uranium mineralization, uraninite, coffinite, thorite, synchysite-(Ce), geochemistry, bohemian massif, central European Variscides

1. Introduction

The Bohemian Massif as a part of the European Variscan belt hosts a significant quantity of different uranium hydrothermal deposits, as well as, vein-type deposits (Aue-Oberschlema, Příbram) and deposits coupled with brittle shear zones (Rožná, Zadní Chodov, Okrouhlá Radouň) [1–4]. The small uranium deposit Tasov, occurred in the north-eastern edge of the Třebíč pluton, described in a recent paper is good example of uranium deposit evolved in brittle shear zones. This uranium deposit was founded during detailed mineral exploration, performed by the Czechoslovak uranium industry (recently DIAMO) in years 1964–1990 [5]. The main aim of presented paper is detailed petrology and geochemistry of unaltered and altered igneous rocks in the north-eastern part of the Třebíč pluton, together with detailed mineralogy of uraninite, coffinite, thorite and synchysite-(Ce) in the area of the Tasov uranium deposit.

2. Geological setting

The nearly triangular Třebíč pluton intruded the boundary between the Drosendorf and Gföhl assemblages of the eastern Moldanubian Zone of the Bohemian Massif (**Figures 1** and **2**) [6]. This pluton was emplaced into a partly exhumed high-grade metamorphic complex of the Moldanubian Zone. The north-eastern contact of this pluton with occurrence of ore mineralization in the Tasov uranium deposit is mostly steep and discordant. The Třebíč pluton is composed by several textural and mineralogical varieties of coarse-grained amphibole-biotite melasyenites to mela-granites, which are usually described as part of the durbachite rock series.

The durbachite rock series represent ultrapotassic (ultra-K) plutonic rocks with highly significant K-feldspar (Kfs) phenocrysts contained in pitch-black, biotite-amphibole-rich groundmass. This rock-series forms a very significant petrogenetic member of the Variscan intrusive rock series of the Variscan orogenic root in the Western and Central Europe [6, 7]. In the Bohemian Massif, the rocks of the durbachite rock series are arranged along two parallel, NNE–SSW (north-northeast-south-southwest) trending belts througth the Bohemian Massif. The Třebíč pluton is dissected by the W-E (West–East) trending Třebíč fault into northern and southern segments. This pluton is, according to detailed geophysical data, interpreted as a relatively thin tabular body currently reaching the depth of aprroximately 2 km. According to these data, the recent Třebíč pluton represents only the erosional relics of large, flat tabular intrusions of ultrapotassic rock series (durbachites s.s.). The north-eastern part of the Třebíč pluton according to detailed geophysical research could be interpreted as a bottom part of this magmatic body [8].

The small occurrence of uranium mineralization by Tasov in the north-eastern edge of the Třebíč pluton represents a very small uranium deposit in the highly important Western Moravian uranium province, which is represented especially by Rožná uranium deposit. All these uranium deposits are coupled with brittle

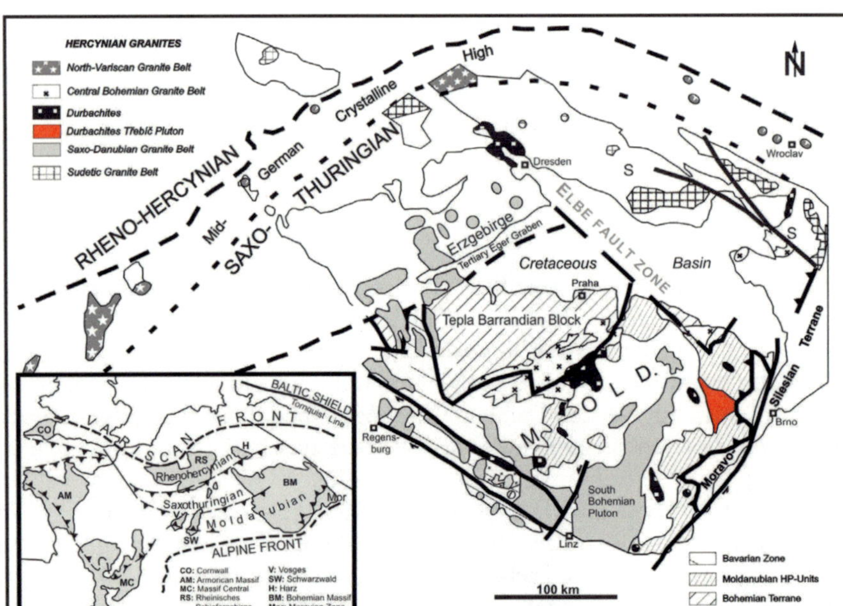

Figure 1.
Occurrence of durbachites in the Bohemian Massif (after Ref. [7] modified by the author).

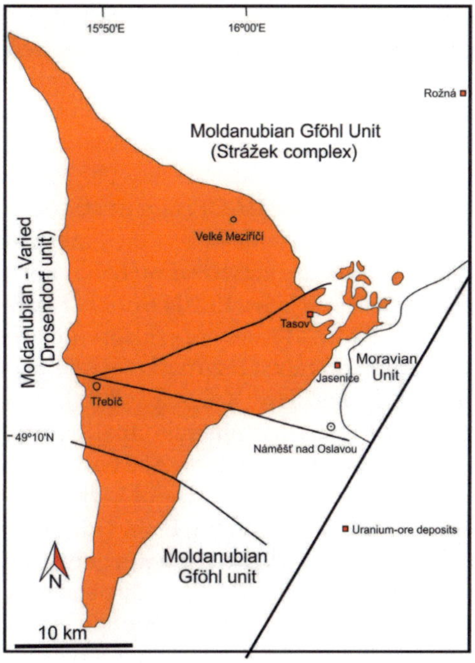

Figure 2.
Schematic geological map of the Třebíč pluton (after Ref. [8] modified by the author).

shear zones. The Rožná uranium deposit is the most significant part of the Rožná-Olší ore district. The uranium mineralization evolved in this ore district is formed by disseminated coffinite-uraninite-U-Zr-silicate mineralization evolved in hydrothermally altered metasediments of the Gföhl unit. This mineralization is concentrated in N-S (north–south) to NNW-SEE (north–north-west-south-southeast) ductile shear zones. The total mine production of this ore district, mined from 1957 to 2017, was 25.142 t of uranium, with an average grade of 0.24 wt. % U [9]. Distinctly smaller uranium deposit Jasenice is coupled with N-S shear zones and is formed by uraninite-coffinite mineralization evolved also in hydrothermally altered metasediments. This uranium deposit is associated with a very complicated system of disjunctive folds. The Early Variscan calcite-uraninite-coffinite mineralization forms the most significant part of this hydrothermal mineralization. The later, Neoidic mineralization is characterized by a very rare occurrence of the vanadium minerals (montroseite, roscoelite), associated with younger generation of uraninite [10]. This uranium deposit was mined in the two periods, namely in 1963–1967 and in 1974–1991, with the total output of 311.2 t of uranium [9]. The small uranium deposit Tasov was mined only very shortly, with a total output of 0.2 t of uranium [9]. However, the total ore reserves of the Tasov uranium ore deposit were estimated as 1200 t of uranium [5].

3. Materials and methods

Detailed petrological and geochemical study of ore mineralization from the Tasov uranium deposit was carried out on a representative suite of the 23 rock samples from selected exploration boreholes (J-S-1, Je-167, J-209), performed by the Czechoslovak

Uranium Industry (recently DIAMO) during their exploration activity in the area between uranium deposits Tasov and Jasenice on the north-eastern edge of the Třebíč pluton. These samples represent unaltered, as well as, altered magmatic rocks of the Třebíč pluton. The major elements were determined by X-ray fluorescence (XRF) spectrometry using the Philips PW 1410 spectrometer at the Central Analytical Laboratory of the Czechoslovak Uranium Industry at Stráž under Ralsko. The iron oxide (FeO) content was measured via titration, whereas the water (H_2O) content was determined gravimetrically. Selected trace elements (REE, U, Th) were determined by inductively coupled plasma mass spectrometry (ICP-MS) techniques at Activation Laboratories Ltd., Ancaster, Ontario, Canada, using a Perkin-Elmer Sciex ELAN 6100 ICP-MS (mass spectrometer), following standard lithium metaborate/tetraborate fusion and acid decomposition of the sample. All analyses were calibrated against international reference materials.

The rock-forming minerals (plagioclase, K-feldspar, biotite), ore minerals (thorite, coffinite, uraninite) and selected accessory minerals (apatite, zircon, allanite, titanite, monazite) were analyzed in polished thin sections. The back-scattered electron (BSE) images were acquired to study the internal structure of individual mineral aggregates. Element abundances of Al, Ca, Ce, Cl, Dy, Er, Eu, F, Fe, Gd, Ho, La, Lu, Mg, Mn, Na, Nd, P, Pb, Pr, Sc, Si, Sm, Sr., Tb, Th, Ti, Tm, U, Y and Yb were determined using a CAMECA SX-100 electron microprobe operated in wavelength-dispersive mode at the Department of Geological Sciences of the Masaryk University in Brno. The concentrations of these elements were determined using an accelerating voltage and beam current of 15 kV and 20 or 40 nA, respectively, with a beam diameter of 2–5 μm. The peak count time was 20 s, and the background time was 10 s for major elements. For the trace elements, the times were 40–60 s on the peaks and 20–40 s on the background positions. The following standards, X-ray lines and crystals (in parentheses) were used: AlK_α-sanidine (TAP), CaK_α-fluorapatite (PET), CeL_α-$CePO_4$ (PET), ClK_α-vanadinite (PET), DyL_α-$DyPO_4$ (LiF), ErL_α-$ErPO_4$ (PET), EuL_β-$EuPO_4$ (LIF), FK_α-topaz (PC1), FeK_α-almandine (LiF), GdL_β-$GdPO_4$ (LiF), HoL_β-$HoPO_4$ (LiF), LaL_α-$LaPO_4$ (PET), LuM_β -LuAg (TAP), MgK_α-spessartine (LIF), MnK_α-spessartine (LLIF), NaK_α-albite (TAP), NdL_β-$NdPO_4$ (LIF), PK_α-fluorapatite (PET), PbM_α-vanadinite (PET), PrL_α-$PrPO_4$ (LIF), SrL_α-$SrSO_4$ (TAP), ScK_α-ScP_5O_{14} (PET), SiK_α-sanidine (TAP), SmL_β-$SmPO_4$ (LIF), TbL_α-$TbPO_4$ (LIF), ThM_β-$CaTh(PO_4)_2$ (PET), TiK_α-anatas (PET), TmL_α-$TmPO_4$ (LiF), UM_β-metallic U (PET), YL_α-YPO_4 (PET) and YbL_α-$YbPO_4$ (LLIF). Intra-REE overlaps were partially resolved using L_α and L_β lines. Empirically determined coincidences were applied after analysis: ThM_α on the PbM_α line and ThM_γ on the UM_β line. The raw data were converted into concentrations using appropriate Pochou and Pichoir (PAP)-matrix corrections [11]. The detection limits were approximately 400–500 ppm for yttrium (Y), 500–800 ppm for REE and 600–700 ppm for U and Th.

4. Results

4.1 Petrology

The samples of the Třebíč pluton, which were taken from the exploration borehole J-S-l, represent the coarse-grained, conspicuously Kfs-porphyritic amphibole-biotite melasyenite (durbachite s.s.). Major components of this rock are K-feldspar (41–45 vol. %), plagioclase (An_{31-36}) (14–19 vol. %), biotite (15–20 vol. %), amphibole (actinolite) (4–7 vol. %) and quartz (up to 5 vol. %). Biotite is represented by phlogopite ($Fe/(Fe + Mg) = 0.45–0.46$, $^{IV}Al = 2.02–2.06$ and $Ti = 0.35–0.42$ apfu).

Accessories are represented by apatite, titanite, zircon, allanite, thorite, monazite and very rare xenotime. Apatite occurs usually as euhedral to anhedral rounded grains of variable size (30–400 μm). According to its chemical composition, apatite corresponds to fluorapatite (2.92–3.53 wt. % F) with content of REE up to 0.81 wt. % REE_2O_3. Titanite in analyzed melasyenites occurs usually as euhedral grains up to some millimeters big. Its composition ranges from 91 to 94 mol. % titanite end-member. Analyzed titanites show also some Al + Fe^{3} excess over F. Calculation of hydroxyl (OH) content allows the estimation of (Al + Fe^{3})-OH component in range of 5–8 mol. %. The content of (Al + Fe^{3})-F component ranges from 0.3 to 3 mol. %. Zircon occurs in euhedral to subhedral grains (25–300 μm), sometimes distinctly zoned. The analyzed zircons contain low concentrations of hafnium (Hf) (0.95–2.14 wt. % HfO_2). The proportion of the hafnium end-member, indicated by atomic ratio Hf/(Zr + Hf), varies from 0.009 to 0.021. All analyzed zircons display lower concentrations of U (0.02–0.53 wt. % UO_2) and Th (0.02–0.33 wt. % ThO_2). The relatively abundant allanite contains 32.4–34.2 wt. % SiO_2, 6.1–13.5 wt. % CaO, 5.3–11.5 wt. % FeO, 0.2–0.4 wt. % MnO, 0.1–2.3 wt. % ThO_2 and 19.2–24.5 wt. % REE_2O_3. For all analyzed allanites, the preference of Ce over La and other REE elements is significant, thus the analyzed allanites could be classified as allanite-(Ce). The magmatic thorite, occurred in unaltered melasyenites, is represented by euhedral 10–40 μm grains, containing 55.1–59.4 wt. % ThO_2, 19.8–22.3 wt. % SiO_2, 0.4–1.7 wt. % ZrO_2 and 0.1–0.4 wt. % Y_2O_3. Monazite contains 7.5–10.1 wt. % ThO_2 and 0.14–0.19 wt. % Y_2O_3. Very rare xenotime was found only as very small inclusion in small zircon grain (25 μm).

In altered melasyenites, the biotite is usually altered in chlorite, K-feldspar is partly altered in muscovite and original plagioclases are altered in albite. The altered melasyenites are usually distinctly carbonatized and highly enriched in calcite. The three major stages of alteration and ore mineralization of the melasyenites in the area of the Tasov uranium occurrence can be distinguished, namely the pre-ore, ore and post-ore stages (**Figure 3**). The presented paragenetic sequence of the ore mineralization from the Tasov area is partly schematic. In altered melasyenites, calcites, albites, hematites and REE-fluorocarbonates are dominant mineral phases, whereas thorite, coffinite and uraninite are less abundant.

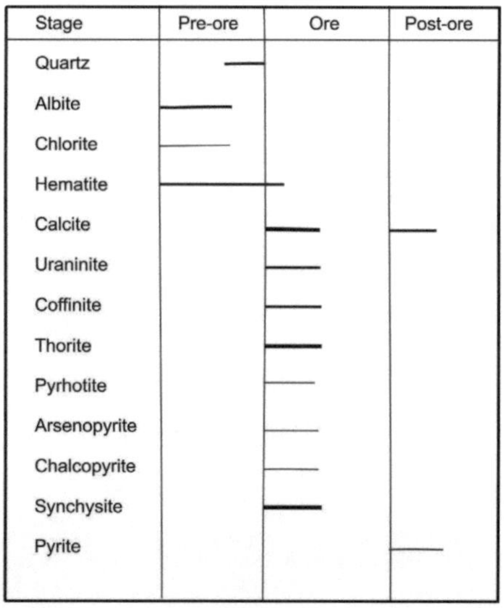

Figure 3.
Paragenetic sequence of the ore mineralization of the Tasov uranium deposit.

4.2 Geochemistry

The studied unaltered amphibole-biotite melasyenites (durbachite s.s.) contain 59–66 wt. % SiO_2 and 3.9–6.9 wt. % MgO. There are ultrapotassic rocks with content of K_2O 5.4–7.0 wt. % and K_2O/Na_2O ratio 2.1–3.7. These melasyenites are slightly metaluminic rocks, with A/CNK ratio (mol. $Al_2O_3/(CaO + Na_2O + K_2O)$ of 0.8–1.1. In addition, high contents of Ba (1200–2060 ppm), Sr. (280–450 ppm) and Th (46–47 ppm) are significant for these rocks (**Table 1**). Chondrite-normalized patterns of analyzed durbachites display high enrichment in light REE with La_N/Yb_N ratio 16.60–18.03 and a moderate Eu anomaly (Eu/Eu* = 0.67–0.78).

For the altered amphibole-biotite melasyenites, the distinctly lower content of Ba (208–827 ppm) and Sr. (63–249 ppm) is highly significant (**Figure 4**). The altered melasyenites are also, in comparison with unaltered melasyenites, enriched in TiO_2 (0.6–3.5 wt. %) versus 0.6–0.9 wt. % TiO_2 in unaltered melasyenites. For hydrothermally altered melasyenites, the high contents of CaO (1.9–4.1 wt. %) and CO_2 (1.2–5.6 wt. %) are also significant (**Table 2**).

4.3 Ore mineralization

The ore mineralization, occurred in area of the Tasov uranium deposit, is highly associated with the NW-SO altered shear zones, which are enriched in albite, calcite and hematite. The ore mineralization is formed by thorite, coffinite and uraninite (**Figure 5, Table 3**).

Sample wt.%	Re-1128	Re-1129	Re-1130	Re-1131
SiO_2	63.74	65.76	61.17	59.18
TiO_2	0.66	0.58	0.80	0.87
Al_2O_3	13.31	13.35	14.09	13.54
Fe_2O_3	2.27	2.49	1.35	1.34
FeO	1.49	0.96	3.16	3.35
MnO	0.05	0.06	0.07	0.08
MgO	4.17	3.49	4.63	5.33
CaO	1.95	1.79	2.61	3.70
Na_2O	2.42	2.49	2.31	1.59
K_2O	6.73	5.92	4.95	5.95
P_2O_5	0.56	0.46	0.66	0.73
CO_2	0.04	0.18	0.73	1.06
L.O.I.	2.00	2.00	3.40	3.50
Total	99.39	99.53	99–93	100.22
A/CNK	0.90	0.97	1.01	0.86
ppm				
Ba	1592	1310	1473	1550
Sr	331	311	334	442

Table 1.
Representative composition of the Třebíč melasyenites.

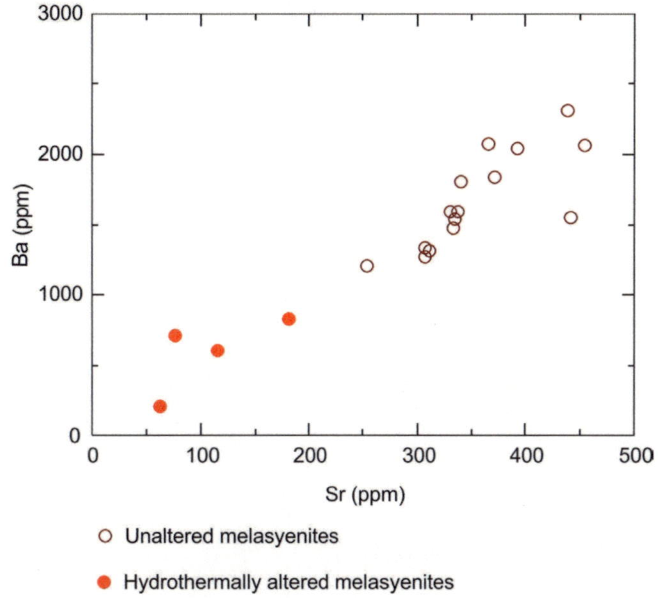

Figure 4.
Distribution of Ba and Sr. in unaltered and hydrothermally altered melasyenites from the Tasov uranium deposit.

Sample wt. %	Re-1198	Re-1203	Re-1208	Re-1209
SiO_2	60.03	44.86	58.48	47.72
TiO_2	0.56	3.48	0.30	3.41
Al_2O_3	11.85	16.01	18.45	13.56
Fe_2O_3	2.28	2.00	0.36	2.17
FeO	0.94	9.19	1.43	9.81
MnO	0.09	0.10	0.03	0.18
MgO	1.66	3.86	0.58	6.98
CaO	7.23	6.06	3.66	6.29
Na_2O	3.10	1.14	4.13	0.40
K_2O	4.40	2.59	7.62	3.48
P_2O_5	0.39	0.52	0.23	0.48
CO_2	2.58	3.78	2.64	1.21
L.O.I.	4.90	6.70	2.10	4.10
Total	100.01	100.29	100.01	99.79
A/CNK	0.52	1.02	0.85	0.87
ppm				
Ba	827	208	604	712
Sr	182	63	116	76

Table 2.
Representative composition of the hydrothermally altered melasyenites from the area of the Tasov uranium ore deposit.

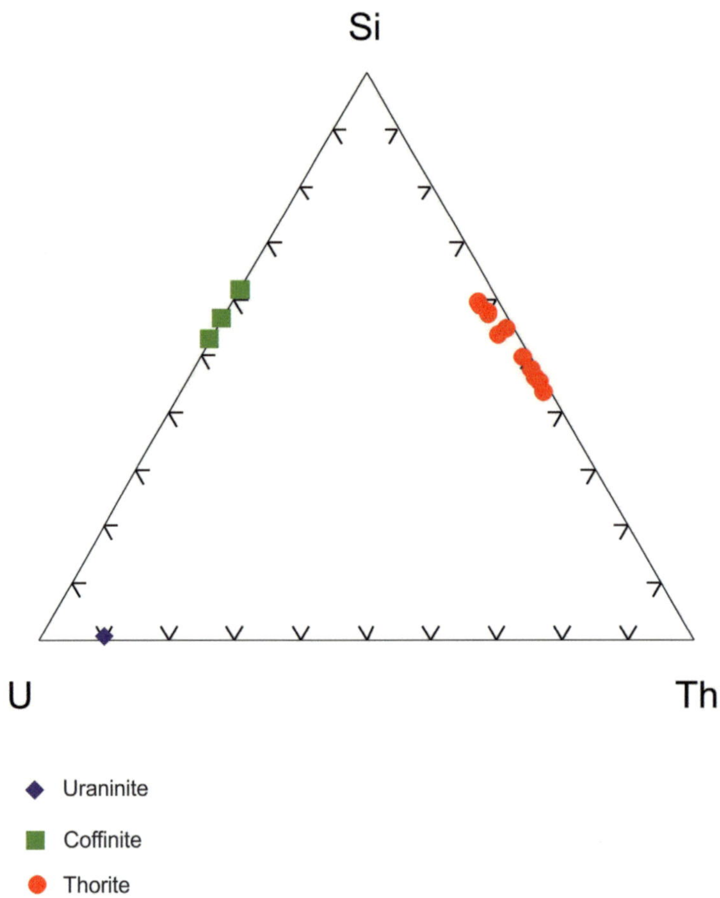

Figure 5.
Composition of uraninite, coffinite and thorite from the Tasov uranium deposit.

Sample wt. %	Re-937	Re-1200	Re-1200	Re-1198	Re-1200	Re-1200
Mineral	Uraninite	Coffinite	Coffinite	Thorite	Thorite	Thorite
UO_2	84.12	59.61	65.50	1.20	3.63	3.43
ThO_2	8.93	0.11	b.d.l.	59.49	58.05	57.20
TiO_2	n.d.	0.04	0.17	0.05	0.12	1.20
FeO	n.d.	0.10	0.29	0.21	0.54	0.91
CaO	0.09	2.33	2.35	6.11	2.24	1.90
MnO	n.d.	0.04	0.05	0.02	0.01	0.01
SiO_2	0.13	21.55	16.40	11.80	18.96	18.94
ZrO_2	n.d.	058	0.43	1.88	0.77	2.10
Nb_2O_5	n.d.	0.05	0.01	b.d.l.	0.02	0.04
Al_2O_3	n.d.	1.83	1.37	0.19	0.71	0.67
PbO	3.79	0.01	b.d.l.	0.02	0.27	0.09
P_2O_5	b.d.l.	0.49	0.63	7.41	1.52	1.44

Sample wt. %	Re-937	Re-1200	Re-1200	Re-1198	Re-1200	Re-1200
La_2O_3	b.d.l.	0.13	0.21	0.83	0.30	0.14
Ce_2O_3	0.32	1.35	1.17	1.06	1.52	1.59
Gd_2O_3	0.19	0.65	0.46	0.31	0.70	0.50
Dy_2O_3	0.18	0.50	0.19	0.25	0.43	0.25
Er_2O_3	0.12	0.22	0.11	0.01	0.08	0.31
Yb_2O_3	n.d.	0.16	0.16	b.d.l.	0.07	0.09
Y_2O_3	0.63	1.38	1.00	0.52	1.38	1.33
Total	98,50	90,55	90,07	88,68	90,55	90,04

n.d.—not determined, b.d.l.—below detection limit.

Table 3.
Representative analyses of uranium and thorium minerals.

Thorite occurs usually as fine-grained aggregates, closely associated with REE-fluorocarbonate synchysite-(Ce) (**Figure 6**). For analyzed thorite, the variation of the UO_2 content is relatively low (1.0–4.9 wt. %). The concentrations' variability of all other constituents in analyzed hydrothermal thorite is also low: the CaO contents

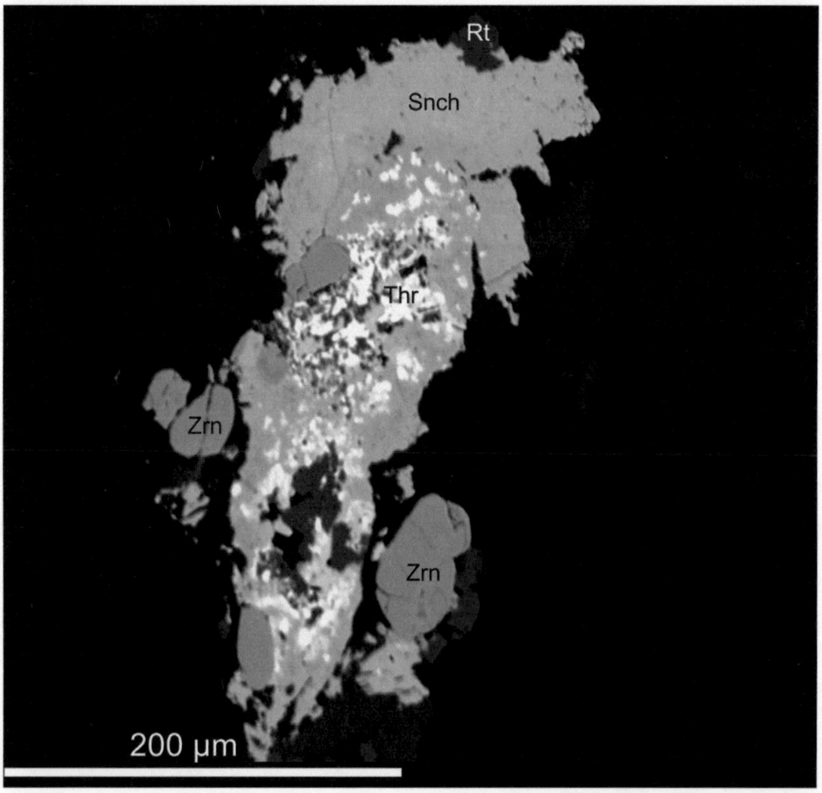

Figure 6.
Back-scattered electron (BSE) images of thorite and synchysite-(Ce) from the Tasov uranium deposit (Rt – rutile, Snch – synchysite-(Ce), Thr – thorite and Zrn – zircon).

Sample wt. %	Re-1198	Re-1208
SO_3	b.d.l.	0.03
P_2O_5	b.d.l.	0.02
CO_2^*	26.26	26.49
SiO_2	0.13	0.13
ThO_2	0.02	0.17
UO_2	0.06	b.d.l.
Y_2O_3	1.05	0.56
La_2O_3	11.84	9.95
Ce_2O_3	21.67	26.07
Pr_2O_3	2.32	2.61
Nd_2O_3	8.79	8.94
Sm_2O_3	1.80	0.95
Eu_2O_3	0.26	b.d.l.
Gd_2O_3	0.80	0.29
Dy_2O_3	0.29	0.07
CaO	16.83	16.99
PbO	0.03	0.02
H_2O^*	0.44	0.62
F	4.77	4.45
O=F	−2.01	−1.87
Total	95.36	96.48
La/Ce	0.55	0.38
La/Nd	..1.35	1.11

*b.d.l.—below detection limit, *) calculated according to stoichiometry.*

Table 4.
Representative analyses of synchysite-(Ce).

vary from 1.7 to 6.1 wt. %, the ZrO_2 from 0.3 to 4.9 wt. %, the FeO from 0.2 to 1.4 wt. % and the P_2O_5 from 1.3 to 6.3 wt. %. Partly higher is variability of the SiO_2 contents, which vary from 10.7 to 19.2 wt. %.

The coffinite occurs usually as small isolated grains, in size of about 10 μm, also associated with synchysite-(Ce). For all analyzed coffinites, there occurs a correlation between SiO_2 and UO_2 with U/Si relations 0.62–0.89. The analyzed coffinites are partly enriched in yttrium (1.0–1.8 wt. % Y_2O_3).

The uraninite was found in mineralized melasyenites as very rare isolated grains in size of about 10–20 μm. The analyzed uraninite is partly enriched in ThO_2 (up to 8.9 wt. %) and Y_2O_3 (up to 0.6 wt. %).

4.4 REE-fluorocarbonates

The cavities and cracks originated during hydrothermal alterations of melasyenites in area of the Tasov uranium deposit were filled by fine-grained aggregates of the REE-fluorocarbonate synchysite-(Ce) (**Figure 6, Table 4**). The analyzed synchysite-(Ce) is light rare earth element (LREE)-dominant with a La/Ce ratio of 0.38–0.55. The concentrations of LREE are 46.4–48.5 wt. %. The content of heavy rare earth elements (HREE) (Gd and Dy) is only 0.4–1.1 wt.%. The contents of Y are relatively low (0.6–1.1 wt. % Y_2O_3). The analyzed synchysite-(Ce) also has a low content of Th (0.02–0.17 wt. % ThO_2).

5. Discussion

5.1 Origin of hydrothermally altered melasyenites

Usually three stages of hydrothermal mineralization could be distinguished in uranium deposits in the Bohemian Massif, associated with brittle shear zones [2, 3, 10]. Usually, the pre-ore, ore and post-ore stages are distinguished. The pre-ore stage is predominantly coupled with occurrence of chlorite and albite, which originated by alteration of biotite and magmatic plagioclases. The strong hydrothermal alteration of original magmatic rocks, enriched in FeO, is accompanied by origin of fine-grained aggregates of hematite. The ore minerals (uraninite, coffinite, thorite) are the most significant minerals of the ore stage. The post-ore stage is characterized by the origin of calcite, some sulphides and REE-fluorocarbonates. However, in the Tasov uranium deposit area the REE-fluorocarbonates originated during the ore stage, together with thorite and calcite.

In the Western Moravian uranium province, which is represented above all uranium mineralization of the Rožná-Olší ore district are three different morphological and mineralogical types of uranium mineralization recognized [12, 13]. The first type of uranium mineralization includes disseminated coffinite-uraninite mineralization occurring in different, usually graphite-enriched shear zones. The second type of uranium mineralization represents vein-type ore accumulations with predominance of uraninite over coffinite in calcite hydrothermal veins. The third type of uranium mineralization represents metasomatic mineralization in hydrothermally highly altered gneisses. For this type, the occurrence of highly hydrothermally altered gneisses is significant. This alteration is characterized predominantly by chloritization, albitization and hematitization of original biotite gneisses, adjacent to shear zones. The uranium mineralization in the area of the Tasov uranium deposit is coupled with this type of uranium mineralization, evolved in the Western Moravian uranium province.

Occurrence of highly hydrothermally altered rocks, coupled with uranium ore deposits in altered gneisses and/or granitic rocks, is significant matter for discussion of their origin. In the past, these altered rocks with highly evolved albitization were named as episyenites. The term episyenites was first used for the description of hydrothermally altered granitic rocks, evolved in metasomatic uranium deposits in the Massif Central and Armorican Massif in France [14, 15]. However, according to the recent international classification of metamorphic and metasomatic rocks, presented and adopted by the International Union of Geological Sciences (IUGS), the original term episyenites could be abandoned and replaced by term aceites [16]. This international commission, which has published the classification of these metasomatic altered rocks, was, however, under high influence of the Russian geoscience school. The recent IUGS classification of hydrothermally altered rocks distinctly delimited the low-temperature hydrothermal alteration (albitization), described as origin of aceites and high-temperature alkali metasomatism as origin of aegirine-amphibole metasomatic rocks and fenites. The term aceite was introduced in geosciences by Omel'yanenko [17]. However, the original literature sources, describing composition and origin of aceites, are available only in Russian [17, 18].

5.2 Sources of thorium and uranium

Uranium and thorium in unaltered melasyenites of the Třebíč pluton are hosted in magmatic accesories, especially in thorite, zircon and allanite. The content of U and Th in ore minerals (thorite, coffinite, uraninite) is most probably coupled with hydrothermal alteration of original accesories. Similar sources of uranium and thorium were proposed also for the Zadní Chodov, Nahošín and Okrouhlá Radouň uranium deposits in the Bohemian Massif [2, 3, 10], which are also coupled with different shear zones.

5.3 Occurrence of thorite

Thorite as a significant ore mineral was described from uranium deposits in the Central Damara Orogen, Namibia, Huangmeijian area, South China, Um Ara area and Abu Hawis area, Eastern Desert area, Egypt [19–22]. All these occurrences represent relatively small uranium-thorium deposits originated by alteration of different granitic rocks.

However, the most significant thorite mineralization in the world that is coupled with U-Cu-REE mineralization occurred in the Singhbhum shear zone (Eastern India) [23] and with Sn-Nb-Ta mineralization in the Madeira (Brazil) [24]. The thorite mineralization in the Singhbhum shear zone is coupled with thorite-bearing bands of different metasediments (e.g., feldspathic, quartz-chlorite, quartz-biotite and muscovite-quartz schists). The thorite-bearing mineralization in the Madeira is coupled with albite-enriched A-type granites and pegmatites. This ore deposit probably represents the largest thorium deposit in the world [24].

5.4 Occurrence and origin of REE-fluorocarbonates

Synchysite-(Ce) is significantly occurring REE-fluorocarbonate, which occurs within different geological environments, especially in carbonatites, pegmatites, altered granitic rocks and in uranium ore deposits. Synchysite-(Ce) was described from the Olympic Dam Fe-Cu-Au-U deposit in Australia and from the Mečichov and Rožná uranium deposits in the Bohemian Massif [25–27]. Occurrences of REE-fluorocarbonates (synchysite-röntgenite) were described also from some small occurrences of metasomatic uranium mineralization in the Třebíč pluton (Velké Meziříčí, Třebíč, Pocoucov, Petrůvky) [12]. According to a detailed study of synchysite-(Ce) from the Olympic Dam uranium deposit, the different origins of REE-fluorocarbonates could be well characterized with the use of La/Ce and La/Nd ratios. The La/Ce ratio for the synchysite-(Ce) from above-mentioned uranium deposits is 0.33–0.59 and the La/Nd ratio is 0.80–1.87. The La/Ce ratio for synchysite-(Ce) from the examined Tasov uranium deposit is 0.38–0.55 and the La/Nd ratio is in the range of 1.11–1.35. These both ratios are very similar with their ratios examined in synchysites-(Ce) from above-mentioned uranium deposits in Australia and the Bohemian Massif.

6. Conclusion

The examined uranium deposit Tasov, which occurred in the north-eastern edge of the Třebíč Pluton, is associated with hydrothermally altered ultrapotassic melasyenites (durbachites). The ore mineralization, which is formed by thorite and coffinite with a relatively highly scarce uraninite, is coupled with highly significant hydrothermal alterations of original melasyenites, which are characterized by the occurrence of albite, chlorite, hematite and a very high amount of calcite, which is closed to the very significant occurrence of REE-fluorocarbonate synchysite-(Ce). The hydrothermal alteration of melasyenites is accompanied by enrichment of Ti, Fe, Ca, and depletion predominantly in Ba and Sr. For the ore mineralization, the occurrence of REE-fluorocarbonate synchysite-(Ce) is highly significant. The La/Ce and La/Nd ratios of the synchysite-(Ce) from the Tasov uranium deposit are very similar to these ratios for synchysite-(Ce) from other uranium ore deposits in the world.

Acknowledgements

The work presented in this paper was supported by the Institutional Research Plan RVO 67985891 of the Institute of Rock Structure and Mechanics of the Czech Academy of Sciences. I am grateful to R. Škoda, R. Čopjaková and J. Haifler from the Department of Geological Sciences of the Masaryk University in Brno for technical assistance by electron microprobe analyses of selected minerals. I am also grateful for constructive comments of an anonymous reviewer.

Conflict of interest

There is no conflict of interest

Author details

Miloš René
Institute of Rock Structure and Mechanics, v.v.i., Czech Academy of Sciences, Prague, Czech Republic

*Address all correspondence to: rene@irsm.cas.cz

References

[1] Schuppan W, Büder W. Lange G on uranium mineralization in the vein deposits of the western Erzgebirge, Germany. Monograph Series on Mineral Deposits. 1994;**31**:191-207

[2] René M. Alteration of granitoids and crystalline rocks and uranium mineralization in the Bor pluton area, bohemian massif, Czech Republic. Ore Geology Review. 2017;**81**:188-200. DOI: 10.1016/j.ore.georev.2016.09.033

[3] René M. Dolníček Z Uraninite, coffinite and brannerite from shear-zone hosted uranium deposits of the bohemian massif (central European Variscan belt). Minerals. 2017;**7**(1):1-17. DOI: 10.3390/min7040050

[4] René M, Dolníček Z, Sejkora J, Škácha P. Šrein V Uraninite, coffinite and ningyoite from vein-type uranium deposits of the bohemian massif (central European Variscan belt). Minerals. 2019;**9**(2):1-21. DOI: 10,3390/min9020123

[5] Hlisnikovský K. Final Report about Results of the Prospecting on the Uranium Ores in the Tasov Area. Unpublished report. Dolní Rožínka: Czechoslovak Uranium Industry; 1993. pp. 1-133. (in Czech)

[6] Janoušek V, Hanžl P, Svojtka M, Hora JM, Erban Kochergina J, Gadas P, et al. Ultrapotassic magmatism in the heyday of the Variscan orogeny: The story of the Třebíč pluton, the largest durbachitic body in the bohemian massif. International Journal of Earth Sciences. 2020;**109**:1767-1810. DOI: 10.1007/s00531-020-01872-2

[7] Finger F, Gerdes A, René M, Riegler G. The Saxo-Danube granite belt: Magmatic response to post-collisional delamination of mantle lithosphere below the south-western sector of the bohemian massif (Variscan orogen). Geologica Carpathica. 2009;**60**:205-212. DOI: 10.2478/v10096-009-0014-3

[8] Leichmann J, Gnojek I, Novák M, Sedlák J, Houzar S. Durbachites from the eastern Moldanubicum (bohemian massif): Erositional relics of large, flat tabular intrusions of ultrapotassic melts–geophysical and petrological record. International Journal of Earth Sciences (Geologische Rundschau). 2016;**106**:59-77. DOI: 10.1007/s00531-016-1296-1

[9] Kafka J, editor. Czech Ore and Uranium Mining Industry. Ostrava: Anagram; 2003. 647 p. (in Czech)

[10] Šouba M, Scharmová M. The character of mineralization of the uranium deposit Jasenice. Sbornik geologických věd. 1984;**26**:131-172. (in Czech)

[11] Pouchou JL, Pichoir F. PAP(φ-ρ-Z) procedure for improved quantitative microanalysis. In: Armstrong JT, editor. Microbeam Analysis. San Francisco: San Francisco Press; 1985. pp. 104-106

[12] Kubeš M, Leichmann J, Wertich V, Čopjaková R, Holá M, Škoda R, et al. Ultrapotassic plutons as a source of uranium vein type U-deposits (Moldanubian zone, bohemian massif): Insights from SIMS uraninite U-Pb dating and trace element geochemistry. Mineralium Deposita. 2024;**xx**:xxi-xxi. DOI: 10.1007/s0126-024—01263-6

[13] Kříbek B, Žák K, Dobeš P, Leichmann J, Pudilová M, René M, et al. The Rožná uranium deposit (bohemian massif, Czech Republic): Shear zone-hosted, late Variscan hydrothermal mineralization. Mineralium Deposita.

2009;**44**:99-128. DOI: 10.1007/s00126-008-0188-0

[14] Cathelineau M. The hydrothermal alkali metasomatism effects on granitic rocks. Quartz dissolution and related subsolidus changes. Journal of Petrology. 1986;**27**:945-965. DOI: 10.1093/petrology127.4.945

[15] Leroy J. The Margnac uranium deposits of the LeCrouzille district (western massif central, France). Geologic and fluid inclusion studies. Economic Geology. 1978;**73**:1611-1634. DOI: 10.2113/gsecongeol.73.8.1811

[16] Fettes D, Desmonds J, editors. Metamorphic Rocks. A Classification and Glossary of Terms. Cambridge: Cambridge University Press; 2007. p. 244

[17] Omel'yanenko BI. Wall-Rock Hydrothermal Alterations. Moscow: Nedra Publishing House; 1978. p. 215. (in Russian)

[18] Zharikov VA, Omel'yanenko BI. Classification of metasomatites. In: Metasomatism and Ore Deposition. Moscow: Nauka Press; 1978. pp. 9-28. (in Russian)

[19] Freemantle GG. Primary uranium mineralization of the Central Damara orogeny, Namibia [thesis]. Johannesburg: University of Witwatersrand; 2015

[20] Zhang L, Wang F, Zhou T, Chen Z, Du X, Zhang S. The origin of uranium deposits related to the Huangmeijian A-type granite from the Lu-Zong volcanic basin, South China: Constraints from zircon U-Pb geochronology and mineral chemistry. Ore Geology Reviews. 2022;**141**:1-17. DOI: 10.1016/j.oregeorev.2021.104665

[21] Ab El-Naby HH. High and low temperature alteration of uranium and thorium minerals, um Ara granites, south Eastern Desert, Egypt. Ore Geology Reviews. 2009;**35**:436-446. DOI: 10.1016/j.oregeorev.2009.02.006

[22] Abdel Gawad AE. Mineral chemistry aspects of radioactive mineralization associated with Zr-, Nb-, and REE-bearing minerals from felsic dikes at Abu Hawis, north Eastern Desert, Egypt. Arabian Journal of Geosciences. 2022;**15**(791):1-19. DOI: 10.1007/s12517-022-10046-0

[23] Sinha DK, Akhila VR, Gupta SC, Mukherjee A, Chakrabarti K, Bhattacharya D, et al. Thorium metallogeny in U-Cu-REE bearing Singhbhum shear zone, Eastern India. Journal of Geological Society of India. 2022;**98**:1051-1057. DOI: 10.1007/s12594-022-2123-x

[24] Hadlich IW, Bastos Neto AC, Botelho NF, Pereira VP. The thorite mineralization in the Madeira Sn-Nb-Ta world-class deposit (Pitinga, Brazil). Ore Geology Reviews. 2019;**105**:445-466. DOI: 10.1016/j.oregeorev.2019.01.004

[25] René M. Alteration of granitoids and uranium mineralization in the Blatná suite of the central bohemian plutonic complex, Czech Republic. Minerals. 2020;**10**(821):1-16. DOI: 10.3390/min10090821

[26] Schmandt DS, Cook NJ, Globanu CJ, Ehrig K, Wade BP, Gilbrt S, et al. Rare earth element fluorocarbonate minerals from the Olympic dam Cu-U-Au-Ag deposit, South Australia. Minerals. 2017;**7**:1-24. DOI: 10.3390/min7100202

[27] Kříbek B, Knésl I, Dobeš P, Veselovský F, Pořádek P, Škoda R, et al. The origin of synchysite-(Ce) and sources of rare earth elements in the Rožná uranium deposit, Czech Republic. Minerals. 2022;**12**:1-26. DOI: 10.3390/min12060690